Perspectives in Biotechnology

NATO ASI Series

Advanced Science Institutes Series

A series presenting the results of activities sponsored by the NATO Science Committee, which aims at the dissemination of advanced scientific and technological knowledge, with a view to strengthening links between scientific communities.

The series is published by an international board of publishers in conjunction with the NATO Scientific Affairs Division

A	Life Sciences	Plenum Publishing Corporation
B	Physics	New York and London
C	Mathematical and Physical Sciences	D. Reidel Publishing Company Dordrecht, Boston, and Lancaster
D	Behavioral and Social Sciences	Martinus Nijhoff Publishers
E	Engineering and Materials Sciences	The Hague, Boston, Dordrecht, and Lancaster
F	Computer and Systems Sciences	Springer-Verlag
G	Ecological Sciences	Berlin, Heidelberg, New York, London,
H	Cell Biology	Paris, and Tokyo

Series A: Life Sciences

Perspectives in Biotechnology

Edited by

J. M. Cardoso Duarte

LNETI
Queluz-de-Baixo, Portugal

L. J. Archer

Gulbenkian Institute of Science
Oeiras, Portugal

and

A. T. Bull and

G. Holt

The Institute for Biotechnological Studies
The University of Kent
Canterbury, England

Springer Science+Business Media, LLC

Proceedings of a NATO Advanced Study Institute on
Recent Developments in Biotechnology,
held March 17–29, 1985,
in Troia, Portugal

Library of Congress Cataloging in Publication Data

NATO Advanced Study Institute on Recent Developments in Biotechnology (1985:
    Troia, Portugal)
    Perspectives in biotechnology.

    (NATO ASI series. Series A, Life sciences; vol. 128)
    "Published in cooperation with NATO Scientific Affairs Division."
    Bibliography: p.
    Includes index.
    1. Biotechnology—Congresses. I. Duarte, J. M. Cardoso (José M. Cardoso) II.
North Atlantic Treaty Organization. Scientific Affairs Division. III. Title. IV. Series:
NATO ASI series. Series A, Life sciences; v. 128.
TP248.14.N38   1985   660′.6   87-11015

ISBN 978-1-4612-9049-0   978-1-4613-1883-5 (eBook)
DOI 10.1007/978-1-4613-1883-5

© 1987 Springer Science+Business Media New York
Originally published by Plenum Press, New York in 1987
Softcover reprint of the hardcover 1st edition 1987

PREFACE

    This book is the proceedings of a NATO Advanced Studies Institute
organized jointly by LNETI, the National Laboratories of the Ministry of
Industry of the Portuguese Government and The Institute for Biotechnological
Studies in the UK. The ASI was held in 1985 on the beautiful peninsula of
Tróia, once the site of a flourishing Roman salt industry.

    The course was the first in the NATO "Double Jump Programme"
specifically aimed to promote industrial and academic participation and
cooperation. As such, contributions across the whole field of biotechnology
were planned and the present volume represents perspectives from specialists
in different areas. Biotechnology has been defined in a recent OECD
publication as "the application of scientific and engineering principles to
the processing of materials by biological agents to provide goods and
services" and the contents of this book, which often describe research from
interdisciplinary groups, reflect this title. The value of the ASI was
further enhanced by many first class poster contributions from the
participants.

    Looking at the development of biotechnology, three generations can be
recognized. The first was based upon the empirical exploitation of microorganisms; the second saw the introduction of scientific and engineering
principles for industrial-scale fermentations; but the third, not
constrained by nor dependent on the scientific experience of the past, is
based on totally novel technology and particularly those of recombinant DNA
methods. We are now embarked on the next major - and possibly last -
industrial revolution. It was against this backcloth that senior academics,
industrialists and many young researchers were brought together to explore
developments in biotechnology - a stimulating experience for all concerned.

    I and my fellow organizers and editors, Professors Archer, Bull and
Holt, would like to express thanks to the Scientific Committee of NATO whose
support made this meeting possible. Our deepest appreciation also goes to
many other organizations and people who contributed to the Institute's
success. There are acknowledgements on the following page. Finally, it is
a pleasure to thank all the course lecturers and participants. I hope this
volume will be a reminder to them of their visit to Portugal and a useful
text to many unable to join us in Tróia.

                                        José C. Duarte
                                        Course Director, Lisbon

ACKNOWLEDGEMENTS

The Organizing Committee and the Course Director wish to thank first the NATO Scientific Affairs Division, the Assistant Secretary General for Scientific and Environmental Affairs and Dr. Craig Sinclair, Director of the ASI Programme for making possible the realization of this ASI and the support given throughout its organisation.

The Portuguese Ministry for Industry and Energy, and the Minister Professor Dr. José Veiga Simão for financial support and encouragement.

LNETI, Laboratório Nacional de Engenharia e Tecnologia Industrial, Portugal, by co-sponsoring the meeting, offering its services and financial support.

IBS, The Institute for Biotechnological Studies, United Kingdom, co-sponsor of the meeting, which greatly contributed to its organization.

The British Council for outstanding cooperation.

INIC, The National Research Council of the Education Ministry of Portugal.

INVOTAN, the Portuguese branch of NATO Science Committee.

Calouste Gulbenkian Foundation, Lisbon

DGT, the Portuguese Tourism Office in Lisbon.

Biogen, S.A.

Celltech Limited

Synthelabo

Porton International plc

AIP, the Portuguese Industrial Association

The poeple at the Training Centre of LNETI, Dr. Emília Raposo and Miss Luísa Ramirez, Amélia Nobre and Mafalda Mendes of the Course Secretariat, Miss Eugénia Lisboa (at the IBS Secretariat) for their constant and skillful assistance.

Special thanks are due to Jane Williams for typing all the contributions for camera-ready copy and to Dr. Clare Ferguson for her invaluable editorial help.

The Organizing Committee:

J.C. Duarte
L. Archer
A.T. Bull
G. Holt

CONTENTS

BIOTECHNOLOGY: OPPORTUNITIES AND CONSTRAINTS

Alan T. Bull

The Institute for Biotechnological Studies
Biological Laboratory, University of Kent
Canterbury, Kent CT2 7NJ, UK

DEFINING BIOTECHNOLOGY

During the decade since 1974 more than thirty national and international reports have been made on biotechnology and most of these have attempted to define the subject. We (Bull, Holt and Lilly, 1982) have chosen to define biotechnology as "the application of scientific and engineering principles to the processing of materials by biological agents to provide goods and services". Such a definition is inclusive not only of actual processes in which the biological agent is used but of those processes also concerned with its preparation and with the processing of biological materials resulting from its action.

Despite the plethora of recent discussions it is worth reiterating that biotechnology is not a distinct discipline but a field of activity; neither is it an industry per se but a group of inter-related technologies that are applicable to a wide range of manufacturing and service industries. Whereas the ancient antecedents of biotechnology are well known, the scientific and commercial excitement of the "new biotechnology" has developed with quickening pace since the emergence of biochemical engineering, enzymes as industrial catalysts and, most recently, genetic engineering. Analysis of this new biotechnology reveals many of the problems that characterise the translation of scientific invention to commercial exploitation: technology push or market pull?; access to information and technology transfer; availability of finance; university-industry relations; roles of government; education, training and manpower provision; patent protection.

The basic conclusions presented in the OECD Report "Biotechnology: International Trends and Perspectives" (Bull et al., loc cit.) have remained valid. It "has proved to be remarkably accurate and has had a positive policy impact by alerting governments to problems" (OECD, 1986). Inevitably the benefit of hindsight leads to a reappraisal of certain issues and events since the Report's preparation argue that insufficient stress was placed on specific scientific issues such as plant biochemistry and molecular biology, protein engineering, membrane technology and bioelectronics. Nevertheless, the comprehensive check list of issues presented as a base for strategic planning (Table 1), and subsequently closely echoed by the US Office of Technology Assessment (OTA, 1984), is a useful point of departure.

Table 1. A Checklist for Strategic Planning in Biotechnology

1. Resources
   Raw materials including feedstocks
                              water
                              minerals
                              energy

       Land availability
       Competing technologies
       Manpower

2. Scientific and Technological Infrastructure
       Education
       Training
       Research base
       R & D priorities
       Information transfer

3. Climate for Innovation
       Invention - innovation time lag
       Industrial base
       Competition
       Finance
       Regulations
       Patent laws
       Social acceptability

4. Trading position
       Commodity prices
       Import-export balances, especially for food
       Markets

5. Environmental Considerations
       Land use
       Effluent and waste, its location and management

From Bull et al. (1982)

     Over the present decade the focus on biotechnology necessarily, and
desirably, has shifted from the wonder of the new science to its commercial
exploitation. The commercial realisation of the new biotechnology broadly
has occurred in two phases. First, one that was dominated by small,
narrowly based companies - in terms of the technology offered - that had
"high tech" profiles and that concentrated on R and D rather than bringing
products and processes to the market place. The second, present phase is
much more orientated towards commercialisation where companies realise that,
among other things, long term success will be built upon market-led
developments, command over several or all of the necessary resources, and an
international operation and structure. The NATO Advanced Study Institute,
of which this book is the proceedings, was the first "Double Jump" programme
in biotechnology and as such sought to provide a forum for university-
industry-government discussions. This chapter does not purport to predict
the detailed development of biotechnology. It is intended to provide a
realistic background against which to explore future developments and as
such it is concerned with (1) the relative positions of present day
biotechnology-based industries, (2) the stimuli for innovation, (3) an
analysis of the major industrial sectors, (4) the characteristics of the new
biotechnology, (5) constraints on development, and (6) the determinants of
successful biotechnology.

Attempts at global analysis of national biotechnology statistics have been made (for example, see OTA, 1984) but are outside the scope of this chapter. Instead brief reference will be made to the U.K. position which reveals many features common to industrialised nations.

In 1980, the contribution of the fermentation sector to the total sales of manufactured products in the UK was about 5% (equivalent to ca. £6500m). Although this sector is dominated by traditional drinks and foods, and by waste water treatment, the relative importance of particular fermentation products can best be assessed in terms of balance of trade figures (Dunnill, 1983; Table 2). We highlighted earlier import-export balance as a major determinant in strategic planning (Bull et al., loc cit.); some obvious considerations which emerge for British planning include:

- possible competition for strong traditional products (whisky);

- political restrictions preventing the improvement of trading position in commodities (sugar);

- need to maintain strong R&D effort to hold the present competitive position (antibiotics);

- opportunities for commercial development of patentable, high value-added products whose manufacture will be influenced particularly by genetic engineering (animal proteins).

Dunnill (1983) also draws attention to the price-volume relationship of biological products, the inverse nature of which has been long recognised in other manufacturing industries. Here the initial attraction of low volume, very high price products (particularly in health care) may have to be tempered by considerations of aggressive, world-wide competition (problems of patent protection; alternative routes to production; small scale production). On the other hand, innovations in the production of and adding value to commodities, such as baker's yeast and beer are likely to provide significant market opportunities.

Table 2. Net Balance of Trade
for the UK in 1980

	Balance £M
Whisky	+ 741
Antibiotics	+ 120
Confectionery	+ 96
Vaccines	+ 7
Enzymes	+ 1
Lysine	- 1
Insulin	- 6
Glutamic Acid	- 6
Beer	- 17
Molasses	- 34
Wine	- 259
Sugar	- 290

From Dunnill (1983)

The above snapshot leads naturally to a consideration of those factors that influence change and innovation in biotechnology. The reader is referred to an earlier analysis (Bull et al., 1979) for background information while the following is offered as an up-date of generally applicable influences.

Crises And Their Resolution

Food and population: although this is a prevailing factor in many parts of the world, the disastrous soya bean harvest of 1973 in the USA clearly promoted renewed interest in single cell proteins.

Energy: the increase in the price of crude oil during the 1970s catalysed enormous (and often ill-conceived) R&D and manufacturing projects for ethanol and other biofuels and chemical feedstocks.

Environment: the gradual appearance of "green" movements over the past 20 years has influenced thinking on waste management and pollution abatements.

Metals: the real, or artificially manoeuvred, shortage of "strategic" metals used particularly in the defence and electronics industries has focussed growing attention on biohydrometallurgical routes to metal winning.

Crises, however, notwithstanding their magnitude and impact, represent moments of danger or disaster but not infallible guides to the planning of biotechnology. Those crises outlined above largely have been resolved, albeit only partially or imperfectly, by factors outside biotechnology, e.g. developments in traditional agriculture, deep sea oil and gas recovery, and material science; political activity and inactivity.

Scarcity

Scarcity here refers to existing market products, such as amino acids, vitamins, steroids, hormones, enzymes, proteins, that are traditionally obtained from animal or plant sources and where there is a need to increase production. The latter has usually been achieved by developing a microbial process or substitute product.

Novelty

Novelty can be viewed in terms of products, services and technologies. The now classical group of novel biotechnology products are antibiotics and their introduction revolutionized chemical engineering and the industrial manufacture of biological materials. Other products in this context include microbial insecticides and monoclonal antibodies.

The novel technologies that have activated biotechnology over the past two decades are: industrial biocatalysts, heralded by penicillin acylase and glucose isomerase in the early 1970s; monoclonal antibody production; genetic engineering, and, subsequently, protein engineering. In particular genetic engineering or recombinant DNA technology has become a major design input with immense potential. Equally clear (now) is the fact that the full expression of genetic engineering in commercial terms is dependent upon (a) factors such as the development of second generation process technologies (for production, extraction and conformation); (b) on the realisation that the major technical problems have moved from DNA splicing to the maintenance of novel genetic combinations and the fidelity of foreign gene expression and translation; and (c) on the wider impact of the technology, in sectors other than health care, and its public acceptability. But, above all else, genetic engineering represents a dramatic technological discontinuity (Sharp, 1985) in the development of biotechnology.

Patterns of Disease

In industrially developed countries the pattern of human disease changed gradually from one dominated by infectious organisms to one progressively replaced by circulatory diseases, cancer and chronic degenerative diseases. On the other hand, infectious diseases continue to predominate in the Third World. In both cases human health care will remain as a, if not the, major instinct for biotechnological innovation.

Social Changes

Dunnill (1983) provides two interesting pointers to the relationship between social trends and biotechnology, both of which refer to changing lifestyles of women. One is the falling sales of baby foods in the UK due to a reduced birth rate and an increase in breast feeding; the other is the substantially increased demand for convenience foods due in large measure to the increased number of working wives.

Economic Changes

"While today's crisis may be unique in its details, its structured features are not unusual. In the past, analogous circumstances have produced instabilities in the market economy very similar to what we are experiencing today" (Mensch, 1979). Economists have produced various models of long-term trends. One of the most well-known is the Kondratieff-Kuznets-Schumpeter model in which economic trends since the Industrial Revolution in the West are depicted by cycles of prosperity, recession, depression and recovery. Extrapolation of this model brings us to the present point of depression and looked-for recovery. Mensch (loc. cit.) favours a model based on a series of intermittent innovative impulses that describe long periods of growth and relatively short intervals of turbulence. He argues that the cyclic model incorporates a deterministic recurrence of phase transition; his metamorphosis model does not, but instead allows for acceleration and deceleration of change. In times of depression the economy is more receptive of basic technological innovations; although some innovations are more likely to be successful than others, there is no determinism in the process. Many commentators agree that among the rush of basic innovations in the 1980s that will drive economic recovery those based in the biotechnologies will have major impact. In one sense, therefore, biotechnology is being espoused widely by governments, companies and investors almost as an act of faith.

THE MAJOR INDUSTRIAL SECTORS

In this section I wish to indicate the range of manufacturing and service industry activities where biotechnology is having and will have a decisive effect. The analysis is not intended to be either comprehensive or detailed but illustrative of opportunities. I have relied to a large extent on the excellent report of Peter Dunnill and Martin Rudd (1984) for statistics and informed comment and this source is recommended to the reader for further insights.

Chemical

Impact of biotechnology on the chemical industry may come from various directions. One long-term objective (post-petrochemicals) will be the provision of renewable feedstocks via ethanol-ethylene, for example. The industry is also concerned with substitute chemicals, especially those at present derived from petroleum. Among the targets are functional and structural polymers, surfactants, resins and fibres but, because of current great overcapacity, new products will need to possess novel properties (e.g. biodegradability, tissue compatibility) in order to replace existing products and to establish new markets. In general terms, large tonnage for chemical products are unlikely to be replaced by biotechnological alternatives this century. Much more attractive options occur in the speciality chemicals sector where the unrivalled capacity of biological processing to yield chiral compounds and to upgrade cheap materials to high value added products is very evident. In this context biocatalyst developments related to operations under non-physiological conditions (e.g. elevated temperature, low water activity) and to extending enzyme and cell half-lives are exciting much interest.

Health Care

The implications of health care for biotechnology have been alluded to above in reference to changing disease patterns. Without doubt health care can be the signal for some of the most rapid and massive responses for support as shown most recently by cancer and AIDS. Little wonder, therefore, that much of the new biotechnology is devoted to problems and markets in this sector. Among the new products emerging are: animal peptides and proteins (interferons, hormones, blood products); monoclonal antibodies (for drug delivery, diagnosis, purification); DNA probes for detecting inborn metabolic errors; vaccines; biomedical products (in situ biosensors, artificial skin, tissue compatible prostheses). Among traditional pharmaceutical products antibiotics are expected to remain very important. Similarly, novel pharmacological products of microbial origin are being discovered (immune regulators, neurological peptides) as a result of innovative screening procedures. The pharmaceuticals sector is subject to increasing international competition due to new entrants into the market (e.g. China) and the temptation of big companies to relinquish out-of-patent antibiotics with the effect of enabling competitors to develop a large manufacturing base from which to infiltrate more lucrative markets.

Agriculture

As Table 3 suggests agriculture is an ideal and immensely varied sector for biotechnological innovation. Opportunities exist world-wide and not infrequently the basic enabling technology is simple, low cost and effective. Plant breeding and micropropagation via tissue culture, for example, has burgeoned after initial scepticism. The targets here are diverse and related not only to performance in the field but to food processing. For example tomato processing in the USA represents a $500m industry and if the solids content of the fruits can be increased from 5 to 6%, $80m savings can be made annually in operating costs.

Food

In very many countries the food processing industries represent the oldest and largest component of the manufacturing sector but also have one of the lowest records of research investment (less than 0.5% of 1982 sales in the UK). Dunnill & Rudd (1985) express doubts about the rapid introduction of high biotechnology into the food industry because of the

Table 3. Biotechnological Innovations in Agriculture

Plant Breeding: Disease and herbicide resistance
                     Increased climatic and edaphic tolerances
                     Increased lysine soya
                     High polyunsaturated oils

Improved Plant Growth: Soil inoculants
                         Growth regulators

Pest Control: Microbial insecticides
                     Microbial herbicides

Animal Breeding: Disease resistance
                         Weight gain

Improved Animal Growth: Bovine growth hormone
                         Feed additives

Silage inoculants

Vaccines: Foot-and-mouth, colibacillosis

inadequate understanding of food structure and its relation to organoleptic
properties. They believe that more immediate gains will come from
innovations in process engineering and preservation. Other biotechnological
goals of the food industry include:

(i)   alternative and novel food ingredients: microbial polymers,
       colouring and flavouring agents, sweeteners, proteins, vitamins;

(ii)   biocatalysis for the modification of food ingredients: high
       fructose syrups, interesterification of fats, detoxification;
       biocatalysis as replacements of traditional materials, e.g.
       rennin, meat tenderizing.

Service Industries

    Under this heading will be considered waste management, energy, and
services per se.

    (i) Waste Management. Because of the very large volumes of very low
value material involved, process costs have to be kept as low as possible
and plant needs to be simple, reliable and long-lived. Hence, there are few
major financial incentives for introducing high technology into this
industry. The main effort has been made in improving existing systems via
better process control, on-line monitoring and instrumentation. Biological
developments are likely to focus on specific problems of toxic waste or
social and operational inconvenience (e.g. deodourisation, clearance of
pipelines and conduits) where inoculation of selected strains and the
maintenance of controlled operational conditions may be important. Although
many countries incur high annual costs for sludge disposal, the alternative
economics of sludge utilisation have often been found wanting. The
economics of product recovery or conversion are very susceptible to world
market prices and the opportunities for utilisation usually depend on sludge
quality, especially metal and phosphate contents.

    (ii) Energy. The present author is among those who take a sceptical
view of the application of biotechnology to energy production. There are,
of course, local conditions and economic circumstances wherein fuels-from-

7

biomass projects appear to be viable (but see Bull, Holt and Lilly, 1982). The claims for biotechnologically enhanced oil recovery, on the other hand, have at best been unduly optimistic and at worst dubious. Finally, Dunnill and Rudd (1985) point to energy-saving contributions of biotechnology in terms of reactions (e.g. use of enzyme catalysis) and downstream processing (separation and purification).

(iii) Services. This is a diverse industrial sector that includes some large activities (toxicity testing; biodeterioration and preservation).

CHARACTERISTICS OF THE NEW BIOTECHNOLOGY

In a recent industrial adjustment and policy report Margaret Sharp (1985) has highlighted three characteristics of the new biotechnology which impinge on policy issues, namely: science-based; technological discontinuity; and prevailing uncertainty. I shall examine each of these features briefly as a forerunner of the final discussion on constraints and determinants of success.

Science-based

Biotechnology is founded in basic science and the driving force is likely to remain with university and research institute activities. Sharp's pithy conclusion is obvious but needs frequent reiteration: "A science-based activity depends upon a core of excellence in science itself. If this is lost, so too is the comparative advantage".

Technological Discontinuity

It is common, and not unreasonable, to delineate three generations of biotechnology. The first, exemplified by traditional fermentation processes, was based upon the empirical exploitation of microorganisms; the second saw the introduction of scientific and engineering principles for initiating and optimizing industrial-scale fermentations; but the third, not constrained by nor dependent on the scientific experience of the past, is based on totally novel technology and particularly that of recombinant DNA manipulation. Sharp's concept of technological discontinuity mirrors Mensch's metamorphosis model of economic trends: both exhibit sigmoidal, overlapping patterns of diffusion and performance, early developments are relatively slow and expensive, but once the point of fast market growth is reached, unit costs fall rapidly until the market is saturated, growth slows and, ultimately, new technologies appear and dominate. A particularly powerful attribute of genetic engineering is that it represents generic technology; not only is it discontinuous but it has major, widespread influences.

Uncertainty

Sharp argues that technological discontinuity creates uncertainty. Do new products have their markets? How big are those markets? Are new production operations economically feasible? What are the lead times to convert basic inventions to market innovations? Such uncertainty about new technology confronts both companies and governments and strategies have varied considerably from country to country. While enabling strategies have been developed by several governments, others have taken more direct action to attack specific, long-term commercial objectives (the French filiere action, for example). In some countries, pre-eminently the United States, venture capital supported start-up companies have pioneered the new technologies and worked on the basis of high risks being rewarded by high profits. Such activity has frequently been followed by comparable in-house developments within the large companies and corporations.

CONSTRAINTS ON DEVELOPMENT

The considerations which need to be made in assessing the risks of entering biotechnology are not unique to biotechnology. Although such considerations are posed here in the form of constraints they are meant to act in a positive sense to aid investment decisions: they are largely self-evident.

(i)   Financial resources: biotechnology is expensive in all but relatively few low-technology areas.

(ii) Technology-push: successful commercialisation is dependent on market-pull and sound judgement on targets.

(iii) Neglect of strategic research and development: biotechnology is a science-based activity.

(iv) Technology transfer: it follows from (iii) that information flow from academe to industry has to be timely and effective.

(v)   Cost of introducing new products: this is especially critical in the health care field where toxicological testing can account for a disproportionately high percentage of total costs (up to 70%) and where the risks of litigation are highest.

(vi) International competition.

(vii) Raw materials availability and costs.

(viii) Competing technologies: biotechnology does not exist in protective isolation. Ignorance of developments in chemical catalysis, material sciences, synthetic drugs, alternative forms of energy and much else will bring the inevitable consequences.

(ix) Patents: the previously defensible distinction between patentable inventions and non-patentable discoveries has become blurred in biotechnology while an added complication stems from the wide divergence of national patent laws relating to biotechnology. The OECD (Beier, Crespi and Straus, 1985) has made a number of recommendations intended to harmonize inter-national patent issues for biotechnology but until these are implemented individual companies and persons will exercise different judgements on choice of patenting and secrecy.

DETERMINANTS OF SUCCESS

Biotechnology like all other innovative industrial developments presents industry and government alike with a "you are damned if you do and damned if you don't" dilemma. The reader is again referred to Margaret Sharp's (1985) analysis for details but her conclusion is simple and irrefutable: "strategies adopted by firms and governments will obviously reflect their own (subjective) judgement of probabilities and attempts to hedge their positions. Not surprisingly, there is considerable variation in strategy". In this case our discussion comes full circle and we can return to the review of strategic planning which opened this chapter (Bull et al., 1982; OTA, 1984). The essential recommendations for governments are facile and well-rehearsed: support the research infrastructure, link industry to the basic sciences, promote commercial applications, encourage joint ventures and licensing arrangements; the challenge comes with the implementation. Following the excessive hype of five to ten years ago, biotechnology has been viewed more cautiously in terms of the next major - and possibly last - industrial revolution. Such reappraisal has been necessary but it has, in a considered way, restated the excitement and real commercial and social benefits that it provides. All those of us engaged in biotechnology could do worse than echo Arthur Dent's reactions to this first sight of "The Hitch Hiker's Guide to the Galaxy": "I like the cover, he said. Don't Panic. It's the first helpful or intelligible thing anybody's said to me all day" (Adams, 1979). The very diversity of biotechnology provides and enormous choice of action, action that should be determined by a company's, an individual's or a government's comparative advantage. As Sharp reminds us, in the long run benefit is derived from use not from discovery or invention.

REFERENCES

Adams, Douglas, 1979, "The Hitch Hiker's Guide to the Galaxy", Pan Books Ltd., London.

Beier, F.K., Crespi, R.S. and Straus, J., 1985, "Biotechnology and Patent Protection", OECD, Paris.

Bull, A.T., Ellwood, D.C. and Ratledge, C., 1979, The changing scene in microbial technology, in: "Microbial Technology: Current State, Future Prospects", A.T. Bull, D.C. Ellwood and C. Ratledge, eds., Cambridge University Press, pp 1-28.

Bull, A.T., Holt, G. and Lilly, M.D., 1982, "Biotechnology: International Trends and Perspectives", OECD, Paris.

Dunnill, P., 1983, The future of biotechnology, Biochem. Soc. Symp., 48: 9-23.

Dunnill, P. and Rudd, M., 1984, Biotechnology and British Industry. A report to the Biotechnology Directorate of the Science and Engineering Research Council, Polaris House, North Star Avenue, Swindon SN2 1ET, UK.

Mensch, G, 1979, "Stalemate in Technology. Innovations overcome the Depression", Ballinger Publ. Co., Cambridge Massachusetts.

OECD, 1986, Long-term Economic Impacts of Biotechnology - Progess Report, SPT (86) 3, pp 1-23.

OTA, 1984, Commercial Biotechnology - An International Analysis, Office of Technology Assessment, Washington, D.C., Government Printing Office.

Sharp, Margaret, 1985, The New Biotechnology. European Governments in search of a Strategy. Sussex European Paper No. 15, University of Sussex, Science Policy Research Unit.

IMMOBILIZED CELL SYSTEMS FOR ENERGY PRODUCTION

J.M. Novais

Laboratório de Engenharia Bioquímica
Instituto Superior Técnico, 1000 Lisboa, Portugal

INTRODUCTION

Immobilization of cells opened new fields to the use of microorganisms to perform certain reactions or to synthesize new products. Immobilization can be regarded as a new way of controlling the capacities of the cells and to provide some of the techniques of chemical engineering for obtaining useful biological products.

Basically, the main difference in the application of free cells and immobilized cells is that the former can only be used in classical batch or continuous fermenters while, with immobilized cells, a wide range of different reactor designs can be considered to be at the service of the imagination of biochemical engineers. A cell is, from the engineering point of view, a catalyst that takes part in one or in a series of chemical reactions in a heterogeneous system.

An immobilized cell is a cell which has in some way been confined to a well defined space, called the bioreactor. It is no longer "free" but is rather in a situation in which it can be, within some limits, controlled by the operator. Immobilized cells however are different from heterogeneous chemical catalysts and the distinctions are mainly two-fold. One is that cells will work at low pressures and temperatures - in general near room conditions. The other is that, if cells are alive, they will be able to reproduce so that a positive increase in activity can result, together with an increase in total "biomass" of the system.

Reactors suitable for the use of immobilized cells can be batch or continuous, the choice depending mainly on the technological development and the volume of products wanted. The properties of an immobilized cell reactor will depend on the method and support used for immobilization, the stability and mass transfer characteristics of the immobilized system and, finally, on the design of the reactor itself.

The methods of immobilization and the supports used will not be reviewed, as they have been widely described elsewhere (Kennedy and Cabral, 1983). Other engineering factors will be evaluated here with a particular emphasis on the utilization of immobilized cell systems for energy production. Much work has been published in this field but in general, results were obtained at a research level and, so far, the number of large scale applications is scarce, although a change in this will be very likely in the near future, particularly in what concerns biogas production. Alcoholic fermentation with immobilized cells on commercial scale is probably not far away.

STABILITY; USEFUL LIFE

Immobilization of cells is carried out with the purpose of allowing for its repeated batch utilization or its continuous use in continuous reactors; this will require a certain stability of the immobilized cells. They are used to reach a well-defined purpose and their activity will be measured by the capacity of the system to reach that specific objective.

Due to the characteristics of biological materials and to the complexity of factors and co-factors involved in each mechanism, it is logical to think that activity will decrease with time, except if the cell is in normal conditions of life and reproduction, and suitable nutrients are fed to the culture. Immobilization itself puts the cell in a set of conditions which in principle is not its optimum and usually a reduction of activity with time is seen. As a consequence of this, the life of a reactor containing immobilized cells is limited and can be measured in relation to the activity reduction. It is customary to take a measurement correspondent to the time needed for an activity reduction of 50%

In continuously working immobilized cell systems, it is usual to notice, after some time, a continuous output of free living cells. This fact can lead the system to a situation of stability in which, independently of the immobilized population, there is a stable population in suspension. Then, the immobilized system itself will act as a continuous innoculum and although the resulting system cannot be considered as immobilized in a strict sense, it in fact conforms to the conditions that in practice are needed to get biomass or a certain product. This system is particularly interesting when an intracellular product is wanted; it has to be optimized then, to yield a maximum amount of free cells.

However, the amount of freed cells is generally somewhat restricted and the flow rates used do not allow a significant increase in suspended cells in the reactor. To optimize this situation, the cells should be immobilized in such a way that most of the new cells may be released immediately to suspension.

Naturally, with living cells the concepts of stability are somewhat more elastic, as compared with immobilized enzymes. In the latter case, there is an irreversible tendency for an activity decrease with time. With living cells, the simple cellular reproduction influences favourably the activity stability and an at least temporary increase in the activity may be observed.

MASS-TRANSFER LIMITATIONS IN IMMOBILIZED CELL SYSTEMS

From a very simple point of view, any heterogeneous reaction will involve a factor of mass transfer resistance as reagents have to be transported to the "site" of the reaction and products have to be returned to the bulk of the solution. The facility of getting reagent and active site in contact, is not equal in all cases and depends on certain factors related to the properties of the environment around the active site and its position relative to the exterior. As a consequence, the immobilization of a biocatalyst involves usually an increase in the resistances in relation with the transport of substrate and product. The importance of this increase in resistance depends on the method of immobilization and on the characteristics of each particular system.

The modification of immobilized cell kinetics (Goldstein, 1976) may be ascribed to factors such as:

(a) Conformation and steric effects. These will arise in cases in which, due to immobilization, a cell can be subject to some kind of stress that may even modify its geometry and its behaviour. Particularly, the free external surface may be reduced and exchanges may therefore suffer a reduction as well. In principle, the internal enzymes will not be affected by this effect because they are protected from the outside conditions by the cell wall.

(b) Partitioning effects. These are related to the chemical nature of the support material and are due to electrostatic hydrophobic and hydro- philic interactions betweeen the matrix and the species of low molecular weight present in solution resulting in a modified microenvironment. When the support has an electrical charge, the kinetic behaviour of the immobilized biocatalyst may differ from the free biocatalyst, even in the absence of mass transfer effects.

Due to these differences which are attributed to partitioning effects, the concentrations of charged species, substrates, products, hydrogen ions, etc., in the domain of the biocatalyst, are different from those in the bulk solution. The main consequence of these effects is a shift in the optimum pH, with a displacement of the pH-activity profile of the immobilized cell, towards more alkaline or acidic pH values for negatively or positively charged carriers, respectively.

(c) Internal and external mass transfer effects. The immobilization of a cell at the surface or within a solid matrix, makes mass transfer effects important, as in that case substrate molecules have to diffuse or travel from the bulk solution to the site of the immobilized cell, and vice- versa for the product molecules.

Problems of external diffusion concern the facility with which a molecule of substrate present in the bulk of the solution reaches the outer surface of the carrier across the stagnant liquid films that are present at the surface of any solid and whose thickness decreases with the increase of turbulence of the surrounding liquid. The same thing happens before a molecule of product excreted from a cell at the surface of the carrier reaches the bulk of the solution.

It should be noticed that this problem has also to be taken into consideration when the cell is not acting in a bioconversion reaction but is just excreting an enzyme that will act in solution to convert a substrate. The driving force involved in external diffusion is related to the difference in molecular concentration between surface and the bulk of solution.

The velocity of flow of the substrate from the bulk of the solution to the surface, is then given by

$$v_{dif} = k_L \, a \, (S_B - S_s) \hspace{4cm} (1)$$

in which K_L is the mass transfer coefficient

a is the surface area of the particles by unit volume

S_B and S_s are bulk and surface concentrations of the substrate, respectively.

When a reaction takes place at the surface of a solid, the flow of substrate to the catalyst surface and its transformation into products take place consecutively so that when steady-state is reached and maintained, it is obvious that the rate of external mass-transfer of substrate V_{dif} equals its transformation by reaction. For a reaction that follows Michaelis-Menten kinetics, the overall rate of reaction V_{obs} is given by

$$V_{obs} = K_L \ a \ (S_B - S_s) = V_{max} \ S_s / (K_m + S_s) \tag{2}$$

To solve this equation, it is necessary to know the values of $K_L a$ and the kinetic constant; otherwise S_s may be graphically obtained by a dimensionless equation:

$$\frac{V_{obs}}{V_{max}} = \frac{B^{-\beta_s}}{\mu} = \frac{\beta_s}{1+\beta_s} \tag{3}$$

where β is the dimensionless concentration of the substrate

$$\beta = \frac{S}{K_m} \tag{4}$$

β_B and β_s are the dimensionless concentrations of the substrate in solution and at the surface of the biocatalyst, respectively.

μ is the dimensionless substrate modulus

The external mass transfer effects on the activity of the immobilized biocatalyst may be expressed by the factor of effectiveness η defined by the ratio between the observed reaction velocity and the kinetic velocity

$$\eta = \frac{V_{obs}}{V_{kin}} \tag{5}$$

If a cell is immobilized within a porous support, another factor has to be considered, namely internal diffusion. Internal diffusion effects are those concerned with the transport of the substrate along the pores of the carrier to reach the cell site and of the products to come back to solution. Internal diffusion is therefore an additional factor beyond the external diffusion already described. Naturally, internal diffusion will depend on the porous structure of the matrix - dimension and tortuosity of the pores - and also on the size of the molecules that have to be transported.

The problem is more complicated when a gas is involved in the reaction. This will, in general, be in solution but when it is produced in large quantities - as for instance in ethanol and biogas productions - the gas may accumulate inside the pores in gaseous phase and the diffusion phenomenon will be still more difficult. The same happens when gas solubility is low.

The situation is also different in the cases of life or death of the cell; in the latter case, substrate molecules have to come to the neighbourhood of the cells, and products will be expelled. In the former situation there is also a need for other nutrients to come to the cell, while the products will also be more diversified.

It is also important to evaluate the transport of substrate molecules through the cell wall; it is not known whether these transport conditions

14

are equal for free and immobilized cells. It is thought that the mechanisms of active or facilitated transport are prevalent at least for the molecules of substrate which are transported in such manner in free cells (Venkatsubramanian and Karkare, 1983). If the cells are not alive, conditions of passive transport have to be considered.

Unlike external diffusion, internal mass transfer proceeds in parallel with the reaction and takes into account the depletion of substrate within the pores, with increasing distance from the surface of the cell support. The rate of reaction will also decrease for the same reason. The overall reaction is therefore dependent both on substrate concentration and/or the distance from the outside support surface.

Let us consider that there is an association between reaction and the diffusion process and that we have a steady state in which the velocities of internal diffusion and reaction are similar. Then, the following equation may be written (Engasser and Horvath, 1973):

$$\frac{d^2s}{dx^2} + \frac{p+1}{x} \frac{dS}{dx} = \frac{kin}{D_{eff}} = \frac{V_{max}S}{D_{eff}(K_m + S)} \tag{6}$$

in which

S is the substrate concentration
x is the distance from the outer surface
p is a geometric factor (+1 for spherical pellets, 0 for cylindrical pellets and -1 for rectangular membranes)
D_{eff} is the effective diffusivity of the substrate in the support

$$D_{eff} = \frac{D_{\Sigma}}{\tau} \tag{7}$$

Σ is the void fraction in the matrix
τ is a tortuosity factor
D is the substrate diffusivity

Analytical solutions of these equations are not possible for Michaelis-Menten type reactions. Numerical solutions are then required. It is therefore necessary to write the equations given in terms of dimensionless terms. For a spherical pellet

$$\frac{d^2\beta}{dZ^2} + \frac{2}{Z} \frac{d\beta}{dZ} = 9 \phi^2 \frac{\beta}{1+\beta} \tag{8}$$

with the boundary conditions

$$\beta = \beta_s \text{ for } Z = 1 \tag{9}$$

and

$$\frac{d\beta}{dZ} = 0 \text{ for } Z = 0 \tag{10}$$

β is the dimensionless substrate concentration
Z is the dimensionless position in the porous support

$Z = \frac{x}{R}$, R is the pellet radius

15

$\emptyset$ is the substrate modulus (a modified Thiele modulus) defined by

$$\emptyset = \frac{R}{3} (V_{max}/K_m D_{eff})^{1/2} \qquad (11)$$

Numerical integration yields the effective rate of reaction V_{obs} as a function of the concentration with the modulus $\emptyset$. The same results can also be represented in the form of graphics of effectiveness factor η against the modified Thiele modulus $\emptyset$.

Internal mass-transfer effects can be reduced by decreasing the particle dimensions of the porous support containing the cells. When external and internal diffusion resistances affect the rate of the reaction simultaneously, the relative contributions of each effect must be estimated separately and quantified by the corresponding effectiveness factors. Hence the overall reaction rate is given by

$$V_{obs} = \eta_{ext} \eta_{int} V_{kin} \qquad (12)$$

In view of the effects just outlined, when the kinetic behaviour of the immobilized cell can be controlled by one or more of these effects, it is useful to distinguish between:

(a) Intrinsic parameters - i.e. the kinetic parameters determined for the free cell.

(b) Inherent rate parameters - i.e. the kinetic parameters that are observed with immobilized cells, in the absence of any diffusional effects.

(c) Effective rate parameters - i.e. the kinetic parameters determined for an immobilized cell system when mass-transfer effects are present and operate in the presence or the sphere of partition effects.

REACTORS FOR IMMOBILIZED CELLS

The design of a reactor suitable to work with immobilized cells is certainly an important factor for an economic and rational utilization of that biocatalyst. In fact one reason to immobilize cells is to provide conditions that allow the worker to use reactors other than conventional fermenters, whether batch or continuous, and in the latter case to be able to work without the problems of wash-out.

Immobilized cells become themselves mini-reactors in which a transformation takes place. These mini-reactors have to be arranged in a certain fashion inside an enclosure, the bioreactor. According to the form of this reactor and to the flow regimen of reagents and products, different types of set-ups may be considered. The reaction taking place is essent-ially a catalytic reaction, as the cell will encourage the transformation without being affected by it.

Biological reactors are basically similar to chemical reactors, but reaction environmental conditions are usually different. With biological catalysts, pH, pressure and temperature must always be moderate and near room conditions, while chemical reactions will use high pressures and temperatures to have their yields increased. This fact is certainly favourable to the design and construction of biological reactors. Operation itself is also favoured because of lower energy consumption.

Biological reactors may be classified in different ways: one of the main concerns of the designer is the mode of operation, which can be either batch or continuous. One of the important reasons to immobilize biocatalysts is to be able to use them in continuous reactors but, in some cases, a repetitive use in batch reactors can be more suitable to the small quantities required.

Batch Reactors

When immobilized cells are used in batch reactors, the system has to allow the separation between cells and medium after each batch, this usually being carried out by filtration or centrifugation. A batch reactor is usually a completely mixed tank, and one way of solving the question of separation of the catalyst, is to insert it inside a basket (a wire mesh enclosure which may be connected to the stirrer and whose apertures are small enough to contain the immobilized cell particles, although usually allowing substrate and product molecules, as well as free cells, to flow through) (Cardoso, 1977). A basket reactor is also a system that protects the immobilized cells, as the stirring paddles will not hit the solid particles. On the other hand, the flow of substrate and/or products through the mesh brings further diffusion problems which may result in a lower efficiency of the reactor.

Another form of confining an immobilized cell system to a reactor is to contain it in a column across which total recirculation of the substrate is carried out. Recycling will be stopped when the required conversion degree is reached.

These two types of modifications correspond to two different types of flow patterns. In the first case, the basket is inserted in a completely mixed reactor in which the composition is equal in all the volume.

In the second, a column batch reactor with recirculation, the flow pattern, while of the plug-flow type, provides for a differential pattern reactor, in which the increase in conversion after each passage is very small and of a differential magnitude. In this case, there is not a measurable concentration difference along the length of the column, and therefore the kinetic reactor equations for the completely mixed reactors can be applied. A plug flow situation, in which concentration of product varies along the length of the column can only be obtained if the flow rate - and the velocity of passage - is very small. The solid biocatalyst may be either packed or fluidised in a column reactor.

Continuous Reactor

The two flow patterns referred to above can also be applied in continuous reactors. In the continuous feed stirred tank reactor (CFSTR), composition at the outlet is necessarily - or theoretically - equal to the composition within the reactor. Therefore composition inside must be high in product concentration and low in substrate.

In the plug flow reactor, conversion will be obtained gradually as the substrate progresses in the reactor. If final conversion is not at the required level, a partial recycle may be inserted in the system.

The choice between the two flow patterns is not arbitrary and it should depend on biological, physical and chemical characteristics of the system. With immobilized cells, the system which has been most widely used is the plug flow. If there is inhibition or toxicity caused by the final product, CFSTR may be inadequate because it puts the cells in contact with the outlet conversion. In the production of ethanol, for example, if final concentration is toxic to the cells, a plug flow reactor will be more suitable. On the other hand, if pH control is required, CFSTR will provide an easier operation, as it is easier to add an acid or an alkali and each addition will be immediately mixed over the whole of the contents of the reactor.

The need to use particulate or viscous substrates may determine the use of a CFSTR. Gas production, such as carbon dioxide in the ethanol fermentation, or the need to provide oxygen to the culture, may also be a factor to be taken into consideration in the choice of the flow pattern.

Different reactor types with different behaviours suitable to different conditions, may be considered in the plug flow pattern. In fluidised bed systems, particles are kept in suspension inside the reactor thanks to the upward flow of the substrate medium. Particles are more free and therefore all their external surfaces are available for changes with the substrate. Control of temperature and pH is also more efficient than in packed beds. Certainly fluidised bed reactors may be useful for particulate substrates which would clog packed beds. Substrate preparation may be less important in the case of fluidised beds, which can be an important asset when large volumes are involved.

In fixed beds, particles containing the immobilized cells are fixed or packed inside the reactor and the liquid goes across that bed. In some early applications, this system has been used such as in trickling filters for sewage treatment and the birch shavings reactor used for vinegar production.

This same method has more recently been applied to biogas production (Messing, 1983). These so-called anaerobic filters consist of a plug-flow reactor with a packing material to which a cellular mixed culture becomes attached. The organic substrate passes through the reactor and part of the carbon gives rise to the so-called biogas made up of methane and carbon dioxide. The gas stream is separated from the liquid after leaving the reactor.

A wide range of packing materials has been used such as rocks, plastic materials, fragments of ceramics, polyurethane foams, etc. The retention of the microbial biomass inside the reactor, immobilized at the surface and in the pores of the packing material, allows for a decrease in the hydraulic residence time from a range of 30 days to less than a week and consequently for a decrease in the volume of the reactor and in its initial cost.

An extreme situation used in the case of biogas production is that in which the reactor has a large surface area and the substrate passes through the reactor upwards with partial recycle. In this case, microbial biomass tends to come to the bottom of the reactor and the liquid flows through that biomass blanket. Cells are therefore immobilized inside the reactor but no deliberate means of immobilization is used. This reactor, called the UASB - upflow anaerobic sludge blanket reactor - has been successfully used for biogas production from mostly liquid substrates.

In terms of biogas production, another configuration has been used, that of the fluidised bed. Microbial cultures responsible for anaerobic digestion have been immobilized on sand particles which were fluidised in a reactor. Two sequential fluidised-bed columns have been used for this purpose. In the first one a specific culture provides the acidification stages of digestion, while the second column performs the methanization stage.

A similar configuration is that of the anaerobic attached film expanded bed (AAFEB). It is a continuous upward-fed partially recycled reactor in which the cells are immobilized in particles of 5 to 200 micron diameter. The flow rate is only enough to achieve a 20% expansion of the bed.

The future of biogas may be extremely bright because of the effort that has been devoted in recent years to the design of new and more efficient reactors. At present, smaller reactors can produce much more gas than was the case two decades ago.

It is interesting to note that alcoholic fermentation has been performed in a reactor in which the cells are pelletized and kept immobilized. This is the so-called tower fermenter (Smith and Greenshields, 1973) which has been developed in England in the sixties for the continuous production of beer.

A large number of classical packed and fluidised bed reactors have been used, at least at laboratory scale, for the production of ethanol.

One of those that has been proposed (Fukushima and Hatakeyama, 1983) and that diverges from the common design is the three stage romboid bioreactor. It is packed with biocatalyst particles that produce ethanol continuously in good conditions, and seem suitable to deal with the problems of the coexistence of the gas - liquid - solid phases. Operating at low values of pH, it has operated continuously over several months without contamination problems.

Immobilized cells packed in semipermeable hollow fiber systems (Hopkinson, 1983) provide another configuration of a packed bed reactor which has recently been arousing wide interest for a number of applications. The system permits the passage of the reactant and the product, but not the whole cells. A hollow fiber system is simultaneously a method for immobilizing microbial cells and a reactor to provide for their use. The fibres are made of polysulphone and acrylic polymers and consist of a highly porous and spongy surface with a thickness of 50 to 75 μ.

In a hollow fiber bioreactor, several hundred fine hollow membrane fibers are packed into a suitable holder, called the shell. This shell side is innoculated with growing cells and the medium is fed into the hollow tubes. The substrate diffuses out into the shell side and the product diffuses back into the hollow tubes. Commercially, fibers with 10,000, 50,000, 100,000 Dalton cut-offs and up to 0.3 μ are available.

In a hollow fiber system, besides the cartridge where the reaction takes place, a pump, a reservoir and tubing are needed, to provide the recirculation of the medium. This system can increase by a factor of several times the cellular density, as compared with free cell fermentation.

In fact one of the important advantages of utilizing immobilized cells concerns the increase in productivity by unit volume of the reactor which is possible to get in comparison with conventional fermenters. This is due in part to an increase in cell density and also to the possibility of putting the reactor to work at high flow-rates without the danger of wash-out. There is no proof however of any intrinsic productivity increase being obtained just because of the fact of having the cells immobilized.

Hollow fiber bioreactors have been used for ethanol production (Mehaia and Cheryan, 1984). Cell densities up to 155 g/l may be obtained and productivities as high as 70 g/l/hr may be achieved when fermenting glucose to ethanol. However carbon dioxide venting and membrane plugging are problems still to be solved, especially if the method is to be used on an industrial scale. Good results have, however, been obtained in the laboratory, pointing to a favourable use of the system.

Another way of utilizing hollow fibers in ethanol production is to have cell recycling (Nishizawa et al., 1983). In that case, reaction takes place in a continuous classical fermenter and the overflow is pumped through a hollow fiber unit to continuously separate cells from the soluble products, the former being sent back to the fermentation reactor.

A variation of plug-flow reactors has been the horizontal (Shiotani and Yamane, 1981) or quasi-horizontal (Cabral and Novais, 1984) column reactor, which is of particular interest in systems in which a large production of gas is expected such as in ethanol production. In fixed vertical beds, CO_2 evolution disturbs the bed and its accumulatin tends to disturb the characteristics of the plug flow across the reactor. In horizontal or quasi-horizontal tubular reactors, the tubes are not filled with the biocatalyst. Therefore, the turbulences provoked by the gas do not disturb the functioning of the reactor and the gas will flow near to the superior face of the tube.

Photosynthetic energy reactions may also involve special problems when immobilized microorganisms are used. In this case, the main point to be taken into consideration is that the light has to reach the particular site where each cell is located. Therefore hollow fibers are for instance inadquate as reactors, as are beds of opaque materials. Glass has been used as support, as have natural polysaccharides such as agar or alginates. These are sufficiently transparent to allow the light through. Packed beds of Botryococcus braunii immobilized in an agar matrix have been used for hydrocarbon production (Fernandes and Novais, unpublished results). Agar transparency seemed to be sufficient for a regular growth of that algal culture.

Hydrogen has been considered an attractive fuel source mainly because it is clean. It can be biologically produced either from bacteria or from algae. In anaerobic systems, hydrogen is produced in a first phase of the biogas process production, and as it has been previously shown, immobilized systems may be used (Suzuki and Karube, 1983).

It is however more interesting to consider hydrogen produced by algae. The production of hydrogen from water is certainly exciting and it may have interest in the future. The influence of environmental conditions, including light intensity, on hydrogen production is being studied. The system is in fact complicated and solutions are not yet entirely at hand.

FINAL CONSIDERATIONS

Deliberate immobilization of microbial cells has been done for more than ten years and much of the mystery that may have been associated with it, has been gradually dissipated. Today, the immobilization of cells is a normal technique although its utilization in industry is not yet so common-place.

It has been recognized that, with immobilized cells, high productivities may be obtained but, in the most common microbial processes, no practical applications have been put forward to replace the common fermentations. Possibly the main reason for this fact lies in the economics of the process. Although higher productivities may be obtained by unit volume, the fact is that in general, immobilization costs and the design and construction of new reactor configurations involve high investment. In low cost processes not demanding in matters of sterility or control, it will be difficult to set up immobilized systems providing competitive costs.

In fermentations in which asepsy and reaction control are important the technology of immobilized cells is probably not yet sufficiently developed to provide economic answers. However for small-scale fermentation - such as fine chemicals production - it has been shown that in many cases it is advisable and economical to replace traditional fermentations by conversion with immobilized cells. This has already been done in the case of steroids and of some organic acids productions.

It is a fact that industry has increasingly become more interested in these techniques and many of the developments taking place have not yet reached public knowledge. The signs are that immobilized cell technology will gradually take important steps forward in the direction of its practical application. After all, ten years is not too long a time to revolutionize a technology which is centuries old.

REFERENCES

Cabral, J.M.S. and Novais, J.M., 1984, Proceedings of the Third European Congress on Biotechnology, 2:381.

Cardoso, J.P., 1977, Ph.D thesis, University of Birmingham, U.K.

Engasser, J.M. and Horvath, C., 1973, J. Theor. Biol., 42:437.

Fukushima, S. and Hatakeyama, H., 1983, in: "Energy from Biomass 2nd EC Conference", A. Strub, P. Chartier and G. Schleser, eds., Applied Science Publishers.

Goldstein, L., 1976, in: "Methods in Enzymology", K. Mosbach, ed., vol. 44 Academic Press Inc., p 397.

Hopkinson, J., 1983, in: "Immobilized Cells and Organelles", vol. I, B. Mattiasson, ed., CRC Press Inc., p 89.

Kennedy, J.F., and Cabral, J.M.S., 1983, in: "Applied Biochemistry and Bioengineering", I. Chibata and L.B. Wingard Jr., eds., vol. 4, Academic Press Inc.

Mehaia, M.A., and Cheryan, L., 1984, Appl. Microbiol. Biotechnol., 20:100.

Messing, R.A., 1983, in: "Annual Reports on Fermentation Processes", vol. 6, Academic Press Inc.

Nishizawa, Y., Mitani, Y., Tamai, M. and Nagai, S., 1983, J. Ferment. Technol., 61:599.

Shiotani, T. and Yamane, 1981, European J. Appli. Biotechnol., 13:96.

Smith, E.L. and Greenshields, R.N., 1973, Biotechnol. Bioeng., Symp. No. 4, 519.

Suzuki, S. and Karube, I., 1983, in: "Applied Biochemistry and Bioengineering", vol. 4, I. Chibata and L.B. Wingard, Jr., eds, Academic Press Inc.

Venkatsubramanian, K. and Karkare, S.B., 1983, in: "Immobilized cells and Organelles", vol. II, B. Mattiasson, etc., CRC Press Inc. p 133.

BIOCONVERSIONS IN ORGANIC SOLVENTS

José C. Duarte

LNETI, 2745 Queluz-de-Baixo, Portugal

INTRODUCTION

Microbes and Enzymes in Organic Environments

Organic solvents are playing an increasing role in the study and application of enzymes. A first comprehensive review of the use of enzymes in non-aqueous solvents appeared a few years ago (Butler, 1979). Since then, the high level of sophistication in organic synthesis has caused an increased demand for enzymes due to their unique capacity to catalyse highly regiospecific and selective reactions. Examples are: the oxidation of the primary alcohol group of several cyclohexane substrates which possess both primary and secondary alcohol substituents by NAD-dependent alcohol dehydrogenase (Jones and Goodbrand, 1977); reactions carried out with fermenting microorganisms such as the oxidation of toluene (Gibson et al., 1970) and the hydroxylation of progesterone (Peterson and Murray, 1952). Many compounds of industrial interest possess very complicated structures (Tamm, 1974). They may also be hydrophobic, requiring the use of organic solvents in order to achieve significant solubility. The wider use of enzymes in the organic laboratory has been advocated by several authors (Jones et al., 1976; Johnson, 1978). The use of organic solvents makes it possible to explore new aspects of enzyme reactions so as to reverse the normal direction of enzymatic reactions (Klibanov et al., 1977).

Microorganisms are known that are able to survive even in extreme organic environments of low water activity. This is the case with microorganisms that proliferate at fuel-water interfaces: a considerable number of microbes travelling short distances into the fuel phase (Schwart and Leathen, 1976). Pichia yeast was grown in an emulsion with oil as the continuous phase (Coty et al., 1971); among 43 microorganisms isolated from samples of jet aircraft fuel systems, 5 isolates of the genera Pseudomonas and Hormodendrum were able to use JP-4 fuel as their sole carbon source (Edmonds and Cooney, 1967); there are microorganisms able to degrade the benzene ring completely (Gibson, 1968) and some Pseudomonas sp. can carry TOL plasmids that enable them to utilize toluene and xylene as sole carbon and energy sources (Williams and Worsey, 1976). E. coli is able to grow in the presence of sublethal concentrations of a variety of organic solvents. The maximal concentration in which growth is possible varies from 852 mM in ethylene glycol to only 4.67 mM in toluene (Ingram, 1977). Changes in phospholipid content and composition are observed during growth on organic solvents.

23

This shows that microorganisms have adapted themselves to the presence of organic solvents, and it is therefore to be expected that enzymes can also function in their presence. Peroxidase shows activity even in several completely anhydrous solvent media (Siegel and Roberts, 1968), mainly in protic solvents. The addition of water always resulted in an increase in activity.

SOLVENT CLASSIFICATIONS AND POLARITY PARAMETERS

To be able to choose an organic solvent for use with enzymes one must have an understanding of solvent-solute interactions.

Solvents can be classified in accordance with several criteria: their chemical bonds, acid-base properties, physical constants or specific solute-solvent interactions. Some classifications have used two or more of these criteria together: dielectric constant values have been used further to subdivide a solvent classification based on the Bronsted-Lowry definition of acids (R. Bates in Coetzee and Richtie, eds., 1969).

Gutmann numbers have been used to measure the strength of Lewis acids and bases. A simple but useful classification (Covington and Dikinson, 1973, p 4) divides organic solvents into:

Protic   - may lose a proton, e.g. alcohols
Aprotic   - may gain a proton, e.g. DMSO
Inert   - such as benzene and hexane

A universal parameter of "polarity" for solvent classifications is hard to define. Dielectric constant and dipole moments have been suggested as a measure of solvent polarity. Solvent polarity is the sum of all specific and non-specific interactions between solvent and solute (Reichardt, 1979a) and one cannot therefore hope that it may be adequately described in terms of individual physical (macroscopic) constants. This leads to the use of empirical scales of solvent polarity, based on convenient, well-known solvent-sensitive reference processes. These scales are limited by the fact that they cannot be used universally, but only for closely related sensitive processes. They are based on linear free-energy relationships (LFE):

$$\log \left(\frac{K_i^B}{K_o^B} \right) = m \log \left(\frac{K_i^A}{K_o^A} \right)$$

where the K's stand for the rate or equilibrium constants for two reaction series (A and B), which are subject to the same changes in the structure of the surrounding medium. K_o^B and K_o^A indicate the reference member of the series. The Donor-Gutmann numbers are an example of these relationships obtained from equilibrium measurements and useful in predicting solute-EPD solvent interactions.

Keto-enol tautomeric equilibria have been useful in establishing polarity scales. An example is the solvent-dependent tautomerization of a pyridoxal 5'-phosphate schiffs' base, which has been claimed to be particularly useful for determining the polarities of protein sites at which pyridoxal 5'-phosphate is bound (Llor and Cortijo, 1977). Kinetic measurements have also been used to establish polarity scales for solvents.

Spectroscopy measurements are the basis of some of the more popular scales. Among them, the Eτ (30), was developed by Dimroth and his collaborators in 1963 (Reichardt, 1979); Eτ (30) values are known for more than 100 pure solvents and for a number of binary mixtures. In accordance with their Eτ (30) values all solvents may be divided into three groups:

Eτ (30) value

	Eτ (30) value
Protic solvents	47 - 63
Dipolar aprotic	40 - 47
Apolar aprotic	30 - 40

Solvents such as formamide (ε = 111.0) and N-methyl formamide (ε = 182.4) are on these scales less polar than water (ε = 78.4). Z and Eτ (30) values have a good linear correlation with the AN values of Gutman, showing that they measure to a large extent the electrophilic properties of solvents.

Another commonly used scale is based on the Hildebrandt and Scott (1949) parameter, δ, defined as the square root of the cohesive energy density divided by the solvent molar volume, $\delta = (E/V_m)^{\frac{1}{2}}$. This value is mainly connected with the energy necessary to form a cavity in the solvent. A good solvent must have a δ value close to that of the solute.

Rohrschneider (1973) measured solvent properties employing gas-liquid chromatographic methods; he used model solutes: ethanol for proton-donor interactions, nitromethane for dipole interactions and dioxane for acceptor interactions. Snyder (1974); 1978) corrected these solubility data for dispersion interactions and molecular weight effects to obtain a solvent polarity scale (P') and selectivity parameters.

The selectivity parameters x_e, x_d, x_n ($x_e + x_d + x_n$ = 1) represent the ability of a given solvent to engage in H-bond or dipole interactions (x_e = proton acceptor; x_d = proton donor; x_n = strong dipole). On this scale solvents are grouped into 8 selectivity classes.

In Table 1 values are shown for polarity parameters for some common solvents.

ORGANIC SOLVENT EFFECTS ON ENZYME STRUCTURE AND ACTIVITY

The effects of organic solvents on enzymes will depend on how much their structure is affected; therefore interactions of solvents with ionic bonds, hydrogen bonding and hydrophobic interactions will determine the response of the enzyme to the use of organic co-solvents. Ionic interactions will be mainly affected by changes in the D.C.; the acid-base characteristics of the solvent will also influence the dissociation constant of the acid or base groups of the protein. Ion pairs may then be formed or broken: strong protic acids (formic or acetic acid) and strong bases (ammonia, pyridine) will have a mainly dissociating effect. Solvents with high H-donating capacity may perturb the H-bond that stabilizes a certain enzyme conformation, due to the relatively weak H-donor group of proteins, -NH; competition with the -C=O proton acceptor group of proteins is more difficult. Completely neutral solvents, like carbon tetrachloride (CT) or benzene, will tend to stabilize H-bonds. The effect of organic solvents on proteins is mainly the result of interference with hydrophobic interactions (Zahler and Niggli, 1977). A number of water-miscible, polar, organic solvents may disrupt hydrophobic interactions (Nishikawa, 1978) by solvating the hydrophobic groups. Herskowitz et al. (1970) established a close relationship between the surface activity of an organic solvent and its denaturing effect on proteins. Poly-alcohols have only a slight effect on surface tension and do not much affect hydrophobic interactions. Other alcohols interact according to the following sequence:

Table 1. Polarity Parameters of Some Common Solvents

Solvent	P'	Rohrschneider Scale[1]			Eτ (30)[2]	Gutmann Numbers[2]	
		x	x	x		DN	AN
Cyclohexane	0.0	-	-	-	31.2	0.0	0.0
Carbon tetrachloride	1.7	0.30	0.38	0.32	32.5	0.0	0.0
Dibutyl ether	1.7	0.53	0.08	0.39	33.4	-	-
Diisopropyl ether	2.2	0.54	0.11	0.35	34.0	-	-
Toluene	2.3	0.32	0.24	0.44	33.9	-	-
1,1,1-Tri chloroethane	-	-	-	-	36.2	-	-
Chlorobenzene	2.7	0.24	0.34	0.42	37.5	-	-
Diethyl ether	2.9	0.55	0.11	0.34	34.6	19.2	3.9
Benzene	3.0	0.29	0.28	0.43	34.5	0.1	8.2
Dibenzyl ether	3.3	0.27	0.27	0.46	-	-	-
Methylene chloride	3.4	0.34	0.17	0.49	41. 1	-	20.4
Ethylene chloride	3.7	0.36	0.19	0.4	41.9	0.0	16.7
Chloroform	4.4	0.28	0.39	0.33	39.1	-	23.1
Ethyl acetate	4.3	0.34	0.25	0.42	38.1	17.1	-
Dioxane	4.8	0.38	0.21	0.41	36.0	14.8	10.8
Acetone	5.4	0.36	0.24	0.40	42.2	17.0	12.5
Ethanol	6.2	0.51	0.21	0.28	51.9	31.5	37.1
Dimethyl formamide	6.4	0.41	0.21	0.38	43.8	26.6	16.0
Dimethyl sulfoxide	6.5	0.35	0.27	0.38	45.0	29.8	19.3
Methanol	6.6	0.51	0.19	0.30	55.5	19.0	41.3
Ethylene glycol	8.5	0.47	0.23	0.30	-	-	-
Water	9.0	0.40	0.34	0.26	63.1	18.0	54.8

[1] In Snyder (1978).
[2] In Reichardt (1979).

methanol<ethanol<2-propanol<t-butanol<n-propanol<2-butanol<n-butanol

The general belief regarding the use of organic solvents with enzymes is that they have inhibitory effects on enzymes. The mechanisms of inhibition are nevertheless not clear. Changes in the dielectric constant of the medium, pKa's, conformational changes and denaturation: all have been involved. More specifically, effects such as nucleophilic participation (mainly by alcohols) in the catalytic process and competitive inhibition have also been suggested as responsible for the observed effects; we shall refer below in more detail to some of these. Different degrees of inhibition were found when a variety of organic solvents (0.1M) were used in the reaction media for the deethylation of ethoxycoumarin and for aryl-hydrocarbon by rat hepatic microsomes (Aitio, 1977); inhibition was almost non-existent for methanol and was maximal for 1-butanol (12 to 16% residual activity). Solvent effects were dependent on the type of reaction: ethanol, acetone and acetonitrile have almost no effect on the hydroxy-lation but decrease the deethylation activity significantly. It is suggested that the inhibition is non-competitive in nature.

Travers and Hillaire (1979) found that while the Ca-ATPase activity of Myosin S_1 was decreased exponentially by ethylene glycol concentration, Mg-ATPase activity was not affected, due probably to a more inaccessible active site. Adding protein or increasing enzyme concentration may afford protection against solvent inhibition, as in the case of the inactivation of erythrocyte acetyl cholinesterase by n-propanol (Dawson, 1976); the substrate protects this enzyme against SDS (sodium dodecyl sulphate) inactivation, but slightly increases the rate of inactivation by propanol, which might mean that propanol binds the enzyme to sites remote from the active site. Simple aliphatic alcohols inactivate the enzyme as a function of their hydrophobicity, as measured by their octanol-water partition coefficient. $\Delta^5 \rightarrow^4$-3-ketosteroid isomerase of <u>Pseudomonas testosteroni</u> was inactivated at a concentration of 10% of 1-propanol and tetrahydrofuran (Jones and Gordon, 1973); glycerol, a solvent considered to be an ideal substitute for water for enzymatic reactions (Klibanov et al., 1978) inhibited hydrolysis of several glucosides by β-glucosidase at a concentration of 50% v/v, while other organic solvents showed a stimulatory effect (Ilio et al., 1978); of these, DMSO, 40% v/v, caused a 25-fold increase in the rate of hydrolysis of genistin; this effect was mainly due to the solubilization of genistin, a substrate practically insoluble in water. With other substrates, solvent effects were different, and salicin hydrolysis was inhibited by MDSO, DMF or dioxane. With cellobiose, only dioxane showed a stimulatory effect. Other cases of stimulation of enzyme activity by organic solvents have been reported: human epidermal trans-glutaminase was activated up to 10 times by pre-incubation with 10 to 50% solutions of alcohols and other solvents (e.g. chloroform) (Plishker et al., 1978); β-glucuronidase specific activity increased almost 30-fold in the presence of tert-butyl alcohol (molar fraction, $x_2 = 0.02$) compared with pure water (Tan and Lovrien, 1972). Carbon tetrachloride in water increased β-glucuronidase activity (Sigma Co., 1958); activity of bacterial phosphoenolpyruvate increased 35-fold in the presence of several organic co-solvents (Sanwall et al., 1966); cholesterol oxidase is also stimulated by solvents such as ethanol, acetonitrile and dioxane over a small range of mole fractions of the organic co-solvent (McGuiness et al, 1978); no correlation was found with the dielectric constant of the binary mixtures; instead, changes in water structure were admitted as an explanation of these effects. Organic solvents can also be helpful in enzyme immobilization techniques: the presence of acetone during cross-linking of glucoamylase in a collagen film resulted in a 2.5-fold increase in activity (Yamado and Mitsugi, 1978).

This data requires an interpretation and explanation of the observed effects which could be used as a basis of predictive models. A useful way of measuring the effect of organic solvents on enzymes is to evaluate their effect on the kinetic parameters of enzymatic reactions. The parameters in question are normally

$$\text{"}k_c \text{ or } k_{cat}\text{"} \quad (V_{max} = k_{cat} \, [E] \,) \quad \text{and} \quad \text{"}K_m\text{"}$$

They are taken from the basic equation of Michaelis-Menten enzyme kinetics, for single-substrate reactions, where

$$K_m = \frac{k_2 + k_{-1}}{k_1} \text{ , and when } k_{-1} \gg k_2, \; K_m$$

has practically the same value as

$$\frac{k_{-1}}{k_1} = Ks,$$

which is the dissociation constant of the ES complex. Therefore the Km values are not always an indicator of substrate affinity and care must be taken in their analysis. The best overall index of substrate susceptibility to enzyme catalysis seems to be the specificity constant, kc/km, in which rate and binding effects are normalized (Bender, 1971). For reactions with two or more substrates the kinetic analysis becomes much more complex (Dixon and Webb, 1964) and the km values obtained under steady-state conditions are even less reliable as indicators of substrate affinity. As to inhibitors, information on their potency can also be obtained from kinetic analysis, and inhibition constants, Ki, given a direct measure of the strength of the enzyme inhibitor. We will refer here to some of the main theories and models used for the interpretation and prediction of the effects of organic solvents in enzymology.

The Electrostatic Model

In this model solvent effects must be an expression of their polarity; the electrostatic theory assumes that electrostatic forces are the ones mainly affected through changes in the D.C.; the same theory has been used to explain solvent effects on enzymes (Laidler, 1965); a correlation between the rate constants of the enzyme reaction and the D.C. has been established (Laidler and Bunting, 1973), with the rate constants depending on the inverse of the D.C.; the model allows the calculation of the electrostatic and non-electrostatic contributions on the entropy of activation; therefore conjectures can be made about the mechanisms involved. Clement and Bender (1963) combined D.C. effects with competitive inhibition to account for the action of aprotic solvents in enzymatic reactions. However theories concerning dielectric constant effects cannot explain the general action of organic solvents on enzymes (Myers and Jacoby, 1973; Maurel, 1978).

Weetall and Wann (1976) attributed to solvents with a lower D.C. the greater enhancement of enzyme activity. Crosby and Lienhard (1970) also found that the enzyme pyruvate decarboxylase is about 10 times faster in ethanol than in water; they suggested that one of the contributions of the enzyme is to create a low dielectric medium.

Inhibition Models

Competitive inhibition is the most common explanation for organic solvent effects on enzyme reactions. Svendsen (1971) showed that the binding of alcohols to subtilisin was responsible for its inhibitory effect; this led to the model developed by Ralston (1972), which allows for multiple binding of the alcohol molecules to either substrate or enzyme. For weak binding of the alcohol and with systems far from saturation the equation

$$\frac{d \ln v}{d [A]} = - n K$$

holds, and logarithmic plots of the reaction velocity (v) against the alcohol concentration [A] will yield straight lines, with the slope depending on the type of alcohol. The binding of "hydrophobic" molecules to proteins is not restricted to alcohols. At least 40 sites on bovine serum albumin are capable of binding butane or pentane (Wishnia and Pinder, 1964) and a large number of sites on β-lactoglobulin were able to bind benzene (Mohammadzadeh et al., 1969). Cremonesi (1977) found a competitive type inhibition for the isomerization reaction of Androst-5-en-3,17-dione in the presence of solvents with low water solubility: butyl acetate, diethyl ether and ethyl acetate. However, the pattern of inhibition was not a simple one. Cordone et al. (1979a) reject the hypothesis that the binding of alcohols is responsible for the competitive inhibition of bovine liver β-galactosidase by monohydric alcohols; instead they claim that preferential solvation by alcohols of non-functional forms of the protein through hydrophobic interactions is a more plausible proposition; they found that substrate protection could result from a stabilization effect on the active conformation; k_c would not be altered unless intramolecular interactions within the enzyme substrate complex were affected by the co-solvent; support for this theory comes also from the effect of alcohols on the binding of oxygen to haemoglobin (Cordone et al., 1979b), an enzyme which is known to exist in two forms: an R "Relaxed" form in the presence of oxygen and a T "Tense" form in the absence of oxygen. The R form has a higher hydrophobic area exposed than the T form and is stabilized at higher alcohol concentrations, while the T form, stablized by salt bridges, is favoured by electrostatic factors at low alcohol concentrations. A similar explanation of the dependence of thermodynamic functions on ethylene glycol concentration with respect to the binding of azide ions to ferrihaemoglobin has been given by Anusiem and Oshodi (1978).

The Extraction Model

Canady and co-workers established the basis of an extraction model for interpreting the inhibition of enzymes by hydrocarbon molecules (Miles et al., 1963; Hymes et al., 1965). They found that the free-energy state of enzyme-complex formation with substrates or inhibitors is a linear function of the surface area of the ligand. Enzymes constitute a non-aqueous phase in which the substrate or inhibitor may remain at a lower potential energy level. Hydrocarbons such as benzene, toluene, xylene, pentane and cyclohexane have verified this relationship for the inhibition of α-chymotrypsin (Hymes et al., 1965). This model has been applied to the effect of organic co-solvents (Falcoz-Kelly et al., 1968) on reactions catalyzed by the enzyme Δ^{5-4}-3-oxosteroid isomerase of Pseudomonas testosteroni; the steroid substrate and inhibitors were dissolved with the help of methanol. Increasing methanol concentration caused an increase in the K_m of the substrate and in the inhibition constants K_i; log $1/K_m$ (or log $1/K_i$) was a linear function of methanol concentration, as was log K_p, the partition coefficient of the substrate or inhibitor between the aqueous phase and isoctane. Effects on k_c are more complex: k_c decreases with methanol content when the substrate is androstendione, but shows an increase for 5-

29

estrenedione and 5-pregnedione at lower methanol concentrations. Hydrophobicity of the steroid molecule is important to determine its affinity for the enzyme and the solvent concentration at which inactivation occurs (Jones and Gordon, 1973). The results indicate that the active site of the enzyme is largely non-polar and that substrate protection is more effective for the more hydrophobic substrates. Maurel (1978) also observed that plots of log (K_m/k_c) against ΔG_{tr}, the free energy of transfer of substrate from water to the various aqueous-organic mixtures

$$\Delta G_{tr} = RT \ln \frac{S_{mix}}{S_{H_2O}}$$

are linear for the hydrolysis of N-acetyl-L-tryptophan ethyl ester by α-chymotrypsin, a reaction in which hydrophobic interactions play an important role. Again the K_m rises with increasing solvent concentrations, while changes in k_c are more difficult to predict: it decreases in this reaction but is practically unchanged in the trypsin hydrolysis of benzoyl-L-arginine ethyl ester (in which hydrophobic and electrostatic forces are involved) and in the hydrolysis of RNA by ribonuclease (a reaction predominantly electrostatic in character). It is noteworthy that no correlation is found for changes in K_m and k_c with other solvent parameters, such as the dielectric constant. Changes in k_c are probably due to interference by the organic solvent with hydrophobic interactions within the protein, which can be expected to differ from one enzyme to another; K_m changes reflect modifications in intermolecular interactions between enzymes and substrates in solution.

Another version of this model is the hydrophobic (or solvophobic) theory of Melander and Horvath (1978), which also relates the value of ΔG with the non-polar contact area; this theory has been verified by the results of Canady and Royer (1968) on the inhibition of several enzymes by aromatics and by the inhibition of pepsin by aliphatic alcohols (Tang, 1965). Use of "activity" instead of "concentration" is desirable so as to obtain a more realistic picture of the inhibition effects (Schlusselberg and Paredes, 1975). This is the case of the acetyl cholinesterase inhibition by alcohols (Dawson, 1976) where it may be concluded that in alcohols with more than 3 methyl groups only part of the molecule is bound to the hydrophobic regions, the other part remaining in the solvent. This model would not apply however to substrates that are highly soluble in water (Cordone et al., 1979a).

The Conformational Model

Conformational changes that lead to an increase in the number of exposed active sites have been advocated as a possible explanation of the activation of human epidermal transglutaminase by several organic solvents (Plishker et al., 1978): preincubation of the enzyme with the organic solvents caused an increase in k_c (as much as 10-fold or more), leaving the K_m unaffected. For the carbamoyl phosphate synthase from rat liver, the apparent K_m for $MgATP^{2-}$ decreased from 7.0mM to 0.1mM when the DMSO concentration was increased to 25% (v/v) (Ishida et al., 1977); the K_m for bicarbonate also decreased, while that for glutamine was not significantly affected; the k_c showed a maximum at 7.5% DMSO. Surprisingly its allosteric inhibition by MgT.UTP was reversed in the presence of high concentrations of DMSO, with an obvious stimulating effect at 30% DMSO; this effect was not displayed by the normal enzyme activator 5-phosphori-bosyl 1-pyrophosphate. This complex behaviour must be related to the complexity of the reaction (three substrates) and the enzyme (Walsh, 1979); it is possible that stabilization due to the presence of organic solvents and the modification of the catalytic and regulatory properties of the enzyme may be related to its mode of action in the living cell. Here also alterations of the quaternary structure of the protein may be related to changes in its kinetics. Tan and Lovrien (1972) had already postulated a conformational motility model to explain the action of alcohols and dioxane used as cosolvents on the catalytic action, denaturation and renaturation of enzymes: activity increases (k_c) due to the presence of the organic solvents of 1.5 to 3.0 and even 10 times were observed, while the K_m seemed to be unaltered; with β-glucuronidase a 32-fold rate increase was observed in the presence of 30% (v/v) of tert-butyl alcohol. This model assumes that the breaking down of water cooperativity in the critical mole fraction, 0.01-0.05, of the co-solvent, is responsible for the increased conformational motility of the enzyme. Co-solvent effects would therefore act mainly on the k_c of the catalytic reaction rather than on the binding parameters. This model was also used to explain the kinetics of the denaturation and renaturation of enzymes in the presence of organic solvents. Other workers have also found that some organic co-solvents affect mostly k_c (Sanwall et al., 1966; Dimroth et al, 1970).

IMMOBILIZATION AND USE OF ORGANIC SOLVENTS

The different models and explanations for the observed effects of the organic solvents on enzymes may also be used to account for the effect of immobilization on enzyme activity and stability. Immobilized enzymes were seen to be normally more resistant to the denaturing action of organic solvents than the free enzymes, therefore allowing the use of higher concentrations of organic solvents. Tanizawa and Bender (1974) used concentrations of dioxane up to 95% with α-chymotrypsin bound to porous glass; the k3 (the deacylation rate constant) and K_m of the reaction depend strongly on the organic solvent concentration. Wann and Horvath (1975) compared the action of solvents (acetone, dioxane, acetonitrile) on soluble and immobilized acid phosphatase; they found that the activity of the immobilized enzyme is reduced more slowly than that of the soluble enzyme with increasing concentrations of the organic solvent, particularly for the acetonitrile-water mixtures; the K_m and k_c of the immobilized preparations both decrease in the presence of the organic solvents, with the specific constant (k_c/K_m) practically unaltered. Immobilization of enzymes enhances their stability in the presence of organic solvents, probably by increasing the actual water concentration in the vicinity of the enzyme. Weetal and Wann (1976) reported increases in the k_c of trypsin covalently bound to porous glass in the presence of optimal concentrations of several organic solvents; solvents with lower dielectric constants supported higher rates and had higher optimum solvent concentrations: in 76% ethanol the K_m was 5 to 6 times greater than in water and the k_c was about 2.5 times greater; solvent concentrations higher than 90% could be used. At higher solvent concentrations (90% in isopropanol) the protein must be completely dehydrated (as revealed by fluorescence studies): this reinforces the idea that it is the immobilization that prevents the collapse of the tertiary structure of the protein and allows some activity to be retained even at high solvent concentrations.

Immobilization of the labile horseradish peroxidase led to better activity retention compared with the soluble enzyme (Epton et al., 1979): the increased stability of the immobilized enzyme in the presence of organic solvents made it possible to use it in reactions never tested before, such as the oxidation of ferrocene, a water-insoluble aromatic compound. Preparation of an immobilized enzyme has also been carried out in completely anhydrous media. Bartling et al. (1974a) prepared a water-insoluble form of lysozyme by crosslinking in anhydrous DMF; the Km of the immobilized enzyme increased by 50% and the kc was unaltered.

The immobilized systems are in fact heterogeneous systems, since matrices for enzyme immobilization are normally solid. Heterogeneity can also be conferred by the presence of more than one liquid phase whether the enzyme is immobilized or soluble. This is the case when a water-immiscible solvent is used. Such systems may be desirable when the substrates are so hydrophobic that they cannot be solubilized by the more polar water-soluble solvents; they can also be used when partition of substrates or products between the organic and aqueous phases helps to prevent enzyme inhibition. Changes in kinetic parameters may be linked to local pH conditions, interface and partition effects among others.

The catalysis and kinetics of heterogeneous systems have mainly been studied using adsorbed films at interfaces (Danielli and Davies, 1951; Davies, 1954). Lipolytic enzymes (McLaren and Packer, 1970) show an increase in activity when present at interfaces, as shown by Brockman et al. (1973) with pancreatic lipase; phopholipases were used in inverted micelles in an organic solvent, diethyl ether (Misiorowski and Wells, 1974; Poom and Wells, 1974). A kinetic model for the phospholipase A action was developed by C.Verger et al, (1973). A review on the heterogeneous biocatalysis of pancreatic lipase and colipase was written by Sémériva and Desnuelle (1976 and 1979).

Hydrocarbon fermentations are another example of heterogeneous enzymatic systems. Hydrocarbons can be used for microbial growth as a carbon and energy source (Humphrey, 1967) and several hypotheses have been put forward for the mechanism of hydrocarbon uptake by cells (Nakahara et al., 1977). In these systems physical factors such as oil drop size or interfacial area assume great importance in enzymatic reaction kinetics or in growth kinetics (in the case of fermentations) (Humphrey and Erikson, 1972; Miura, 1978).

Enzymes in Emulsions

Enzymes may be used in heterogeneous organic systems via emulsion formation (May and Li, 1974; Li et al., 1975): an aqueous solution with the enzyme is emulsified with an immiscible liquid containing a surfactant. Additional substrate specificity conferred by the relative solubilities and diffusivities in the hydrocarbon and aqueous phases of the substrate and products was used to oxidize phenol in a mixture of phenol and protocatechuic acid (the acid does not diffuse into the hydrocarbon phase); this technique may be suitable for the immobilization of multicomponent enzyme systems. Leavit et al. (1974) used a similar technique for oxidation of tetradecyl alcohol. Here the organic phase was initially dispersed in the aqueous phase; at oil concentrations higher than 1:2 inversion occurred, causing an increase in the initial rate of oxidation. The use of organic solvents to assist water-insoluble substrate transformations by enzymes was also advocated by Antonini (1975). The organic solvent ensures a virtual saturation concentration of the substrate in the aqueous phase. Continuous extraction of the reaction products also occurs in these systems. Examples are the action of lipase on olive oil dissolved in dichloroethane, the enzymatic conversion of steroids, the oxidation of aromatic hydrocarbons and the conversion of salicylate into catechol when an appropriate organic solvent is used for solubilization of the substrate. Oxidation of butanol by alcohol dehydrogenase occurred without enzyme inactivation when a two phase system consisting of benzene and water was used; concentrations of butanol higher than 3% in the aqueous phase were thus avoided.

This technique was also used for resolution of racemic mixtures (Uzuki et al., 1975). Inverted micelles can be employed to solubilize enzymes for use in apolar solvents (Balny and Douzou, 1979; Wolf and Luigi, 1979) with retention of activity; in this way the amount of water, size of micelles, polarity, and ionic composition of the environment can be more easily controlled. Martinek et al. (1977) used sodium bis (2-ethyl-hexyl sulpho-succinate) as a surfactant to solubilize α-chymotrypsin and peroxidase in inverted micelles with octane and benzene as the apolar solvent. The specific constant (K_c/K_m) of trypsin-catalysed reactions was similar to that of aqueous reactions, and activity could be maintained for months at room temperature; peroxidase would act on a variety of substrates: H_2O_2, ferroxyanide and pyrogallol; the inhibitory effects of pyrogallol could in this way be avoided. A study of the influence of water concentration on phospholipase A_2 solubilized in inverted micelles of phosphatidylcholine in diethyl ether-methanol (95:5 v/v) (Misiorowsky and Wells, 1974) showed the enzyme is active only when present in moderately hydrated micellar species.

Free Cells and Immobilized Enzymes

Nocardia rhodochrous cells with cholesterol oxidase activity were used to convert cholesterol into cholestenone, the first step in the microbial oxidation of cholesterol (Buckland et al., 1975); the substrate cholesterol was dissolved in an organic solvent. The best solvents, which gave higher conversions, were the most apolar: toluene, hexadecane or carbon tetrachloride. Water-miscible solvents did not give higher conversions than water (Table 2). With this system high cell concentrations and high substrate concentrations (in the organic phase) could be used to give extraordinarily high conversion rates. Transformations performed by growing cells are also possible in such systems. This is the case in the conversion of 1,7-octadiene to 7,8-epoxy-1-octene by a Pseudomonas sp. growing in octane (1% v/v) with an organic immiscible phase consisting of cyclohexane, 20-60% v/v (Schwartz and McCoy, 1977); the toxic product could in this way be continuously extracted and 88.5% molar conversion attained in 72 hours compared with 16.7% in the completely aqueous medium. The growth of cells in a water in oil (w/o) emulsion with n-hexadecane as the substrate resulted in a threefold increase in concentration of cells in the aqueous phase (Coty et al., 1971), which would facilitate cell filtration or centrifugation.

Enzymes immobilized in porous materials embedded in water can be dispersed in water-immiscible organic phases containing the substrate (Klibanov et al., 1977): chymotrypsin immobilized in porous glass beads was capable of catalyzing the synthesis of N-acetyl-L-tryptophan (10 M) and ethanol (1M); the reagents were solubilized in chloroform; the conversion was displaced towards the synthesis of the ester with a yield of almost 100% compared with 0.01% for the completely aqueous system. Klibanov and co-workers attributed this equilibrium displacement to two factors: the concentration of water (itself a product of the reaction) is lowered, and there is a change in the true equilibrium constant K, compared with K_w, the constant for the aqueous system, due to the partition effects of substrates and products in the two phases. Chloroform was more effective than benzene or ether, and miscible organic solvents did not even support the hydrolytic activity of the enzyme. Another example is the synthesis of glycerophos-phate by alkaline phosphatase in a similar system with a yield of 30% compared with the value of 0.1% in water. Systems of this kind are useful on lipase applications because of the enzymes' requirements for an interface (Brockman et al., 1973; Sémériva et al., 1974).

CRITERIA FOR USE OF ORGANIC SOLVENTS IN ENZYMATIC REACTIONS

From the above it may be concluded that to chose a priori a solvent to use in an organic aqueous enzymatic system is not an easy task. The use of single physicochemical parameters for the characterization of solvents or

Table 2. Influence of the Nature of the Organic Solvent on Enzymatic Steroid Conversion

Solvent	Reaction Conv. (%) A	B(1)	Solubilities (wt, %) Solvent (in water)	Cholesterol (in solv)	Cortisone (in solv)
n-Hexane	-	>5(98)	0.0009	1.92	0.003
iso-Octane	-	>5(100)	0.0002	-	-
Toluene	100	-	0.054	-	-
Hexadecane	92	-	-	-	-
Carbon Tetrachlor.	87	>5(100)	0.08	14	0.0025
di-Ethyl Ether	75	10(38)	6.9	26	0.024
tri-Chloro Ethylene	-	10(90)	0.1	-	-
Chlorobenzene	-	15(95)	0.048	-	0.027
Benzene	62	-	0.178	14	-
Petroleum Ether 60-80	42	-	-	-	-
Cyclohexane	33	-	0.0055	-	-
Chloroform	33	-	0.82	15	-
Ethyl Acetate	-	90(29)	9.7	-	0.303
Butyl Acetate	-	100(48)	1.0	-	0.115
Acetone	12	-	miscible	4.2	-
Isopropanol	12	-	miscible	-	-
Ethanol	8	-	miscible	1.6	-
Water	12	-(100)	-	1.8*10	-

Reaction A: Cholesterol Oxidation (Buckland et al., 1975).
Reaction B: Cortisone Reduction (Cremonesi et al., 1975).
(1) For reaction B enzyme activity (%) is shown in brackets.

their mixtures, such as the dielectric constant (Laidler, 1965) or the surface tension (Connors and Sun, 1971; Torma, 1976) is insufficient to explain the multitude of effects that the solvents have on enzymes and other biological materials, even if they are of limited validity in particular cases. This failure is related to the absence of any correspondence between macroscopic parameters and the effective microscopic properties of mixtures: these may vary with the solute because of differential solvation, giving rise to microscopic heterogeneities (Douzou, 1977). A description of these microenvironmental effects is not possible at this stage; even in the case of homogeneous chemical reactions we are in the "dark ages" of progress as to the correct answer (Amis, 1973).

The Solvophobic Power

It is accepted (Reichardt, 1979) that the influence of the solvent on reaction rates is determined by the difference in the free energies and in the enthalpies and entropies of solvation of the reactants and transition states. The use of the absolute rate theory relates these changes to the activity coefficient of the species in solution. In enzymology, activity coefficients were used by Schlusselberg and Paredes (1975) because they were more realistic than concentration in giving a picture of the hydrophobic interactions of alcohols with proteins. However, activity coefficient data are not easily available and its generalized use in enzymatic catalysis is not foreseeable in the near future. Melander and Horvath (1977, 1978) tried to explain salt effects on hydrophobic interactions occuring in precipitation and chromatography of proteins, on enzyme inhibition by small molecules of hydrophobic character and on enzyme immobilization on hydrophobic surfaces.

The interactions were described in terms of "non polar surface area" and "surface tension increment". Ray (1971) used surface tension changes in solvents to measure the critical micelle concentration of a detergent in these solvents and classified the solvophobic power of a solvent in accordance with their capacity to form micelles. The solvent solvophobic power did not correlate with Hildebrandt's solubility parameter, or P' (see Table 1), but follows the proton donor (x_d) capacity of the solvent. Solubility parameter mapping was used by Eisenbach et al. (1979) to measure solvent interactions with purple membrane (from Halobacterium halobium), which succeeded in demonstrating the high hydrophobicity of the membrane. Solvophobic solvent power was used as a criterion in the choice of a solvent for enzymatic reactions by Klibanov et al. (1978), who according to Ray (1971) predicted that glycerol and solvents of Class I would be the best substitutes for water in enzymatic reactions, because of their capacity to sustain the hydrophobic interactions of the protein. They verified that 5% of the initial activity of acid phosphatase in the hydrolysis of p-nitrophenyl phosphate was maintained in 86% glycerol compared with 62%DMF (Class II) and 63% ethanol (Class III).

Miscible Versus Immiscible Solvents

One of the first decisions to take when choosing an organic solvent is whether a water miscible or immiscible solvent should be used. Although immiscible organic solvents normally have higher denaturing effects more water miscible solvents have to be used in higher proportions with increasingly hydrophobic substrates. The problem always remains of defining the degree of immiscibility. In any case a balance between activity and stability must be considered as exemplified in experiments by Buckland et al. (1975) and Cremonesi et al. (1975), summarized in Table 2. In the case of reaction B, having NAD as a co-factor, the less the solvent is soluble in water the higher is the enzyme activity (not shown in the table). Carbon tetrachloride seems to support an enzyme activity higher than predicted by its solubility in water when compared with chlorobenzene. However, if one considers the polarity of carbon tetrachloride, P', it is lower than that of chlorobenzene, as it is its electrophilicity as assessed by its $E\tau(30)$ values compared with the relatively high x_d value for chlorobenzene. The low value of the activity in the presence of butyl acetate, despite its relatively low solubility in water, may be due to a high x_e (nucleophilicity) value as compared to the ethyl acetate value; it is also noteworthy that the $E\tau(30)$ value of butyl acetate is high. In reaction A (Buckland et al., 1975), an oxidation reaction not NAD dependent, the enzyme is "immobilized" in the cells. The more water-immiscible solvents give higher conversion rates; they also give higher cholesterol solubility. Substrate solubility is of importance for achieving a high conversion, and considering again reaction B, it is evident that the highest conversion is obtained for butyl acetate, an obvious compromise between enzyme activity and substrate solubility. For reaction system A the high values obtained for ethyl ether and toluene are somewhat more difficult to explain as is the low value for cyclohexane. The higher conversion value of toluene than of n-hexadecane and carbon tetrachloride (CT), both less "polar" than toluene, may be due to its lower water solubility (compared with CT) and to its low electrophobicity value as assessed by the Acceptor Numbers (AN) of Gutmann (Reichardt, 1979).

The value for diethyl ether, with the highest substrate solubility, could again be explained by its very low electrophilicity as assessed by the Acceptor Number, but the contrary low conversion rates obtained with chloroform may be due to high electrophilicity ($x_d = 0.41$); $E\tau(30) = 39.1$). The low value for cyclohexane is more difficult to explain (no value for cholesterol solubility in cyclohexane was obtained); one could argue that cyclohexane has a structure somewhat similar to that of the reactive end of the substrate molecule that binds to the active site, and therefore could be responsible for inhibition of the enzyme reaction.

Another factor to bear in mind is the stability of the enzyme in the presence of interfaces in the heterogeneous system. "Immobilization" in the cells in reaction system A may afford some protection, whereas in reaction system B this effect is probably achieved by the addition of serum albumin to the system (Cremonesi et al., 1975). Susceptibility to interfacial denaturation may be related with the interfacial tension between the phases. Solvents with high surface tensions showed the best conversions (as is the case of toluene). In contrast, water miscible solvents with high polarity and low substrate solubility cannot afford higher conversion rates than water. Duarte and Lilly (1980) tested the cholesterol oxidase activity on two different series of solvents. The highest levels of activity were obtained with chloro-derivatives of ethane and with 6-carbon ethers. The smaller compounds are probably interfering with the binding of cholesterol as was the case for isomerization. Another explanation is that solvation of cholesterol by these solvents leaves the -OH group in the right position for reaction. These interpretations are tentative. Different types of reactions and even enzymes from different sources may prove to be differently affected (Lugaro et al., 1973).

Cremonesi (1977) stated that the best solvent would be the one which had a smaller effect on the K_m of the reaction and at the same time ensured a satisfactory concentration of the substrate in the water phase. However as previously seen consideration has also to be given to factors affecting Kc and the operational stability of the enzyme(s) may also have to be considered.

Reactivity and Stereospecificity

Two other factors to bear in mind when choosing a solvent are its effect on enzyme reactivity/specificity. Specificity may be altered by the use of high solvent percentages. The lower the reactivity of the substrate, the less it is affected by the use of an organic solvent, and the reactivity order can even be reversed (Jones and Gordon, 1973; Klibanov et al., 1978; Jones and Mehes, 1979); in 95% glycerol the order of the rates of hydrolysis, which in completely aqueous systems is p-nitrophenyl phosphate > β-glycerophosphate > α-glycerophosphate, is reversed (Klibanov et al., 1978). The stereospecificity of enzymes is one of their characteristics of extreme importance in asymetric organic synthesis; therefore it is important to know how the stereospecificity is affected by the organic solvent. Jones and Mehes (1979) found that α-chymotrypsin, an enzyme with high stereospecificity, retains its normal L-enantiomeric preference with the enantiomers of methyl 2-acetamide-2-phenyl acetate in the presence of the solvents used. DMSO was the solvent which exerted least influence on enzyme stereospecificity, and dioxane the one which affected it the most. Kinetic parameters of the two enantiomers were also differently affected. However, it seems that at very large cosolvent concentrations stereospecificity cannot be guaranteed; but it is to be expected that at these concentrations loss of activity may even precede the disappearance of stereospecificity; water-immiscible solvents were not tested.

The Use of Solvents to Modify Unfavourable Equilibria

The choice of an organic solvent will depend on the reaction system and on the objectives of the work. It is often desirable to carry enzymatic reactions to completion even when the equilibrium is unfavourable. It should be realized that enzymes serve only to increase the rate at which a particular reaction reaches equilibrium, but they do not influence the free energy of the system. A reaction can be carried to completion, even when the equilibrium constant is not favourable, by continual removal of the desired product. This was applied to the papain catalysis of the formation of amide bonds, reversing the normal hydrolysis (Kirschenbaum, 1963): by using relatively hydrophobic aromatic amines the resulting amide product precipitates from the aqueous reaction medium and the reaction goes on in the synthetic direction until reactants are no longer available.

We may consider the way enzymes in cells work in the synthetic direction and use organic solvents in modelling some of the _in vivo_ reactions. The following possibilities may be considered for the use of solvents to modify unfavourable equilibria:

- Lowering water activity and concentration, when water is one of the products. In this case a water miscible or immiscible may need to be used.

- Functioning as a reservoir of the substrate (water immiscible solvents) or increasing its solubility (water miscible solvents).

- Functioning as a sink of products and avoiding possible inhibitory effects (water immiscible solvents) or increasing their solubility (water miscible solvents).

- Creating interfaces, in the case of interface promoted catalysis (water immiscible solvents, emulsions).

CONCLUSION: COMMERCIAL PERSPECTIVES

Many synthetic enzymatic reactions can be carried out using the techniques referred to herein: synthesis of glycerophosphate (Klibanov et al., 1978); synthesis of esters (Ingalis et al., 1975); sucrose synthesis (Kelly et al., 1976); glyceride synthesis (Iwasaki, 1976); synthesis of urea (Butler and Reithel, 1977), among others already referred to in this chapter. It is reasonable to believe that developments in the use of organic solvents will increase the number of applications of enzymatic reactions in organic synthesis and in industry in general.

In Table 3 some applications of potential industrial value are shown. Some of these processes are already in production, at least on a pilot scale, and indications are that in the relatively near future other applications will find their way into the market.

Table 3. Application of Organic Solvents in Bioconversions

Application	Enzyme	Org. solv.	Ref.
Oxidation of cholesterol	Oxidase (free Noc. cells)	Carbon tetr. and others	Buckland, 1975
Production of D-phenyl glycine	Subtilisin (immob.)	Methylisobutyl ketone	Bayer, 1979
Oxidation of cholesterol	Oxidase (immob. Noc. cells)	Carbon tetr.	Duarte, 1980
L-menthol production	Lipase (entrp. cells)	n-Heptane	Omata, 1981
Synthesis of aspartame	Thermolysin (immob.)	Ethyl acetate	Oyama, 1981
Organic synthesis	HLDA	Water-misc. (protic/aprotic)	Jones, 1982
Oil interesterification	Lipase (immob.)	n-Hexane	Yokozeki, 1982
Cholesterol transformation	Several enzymes (immob. Nocardia)	1,1,1-TCE	Duarte, 1983
Organic synthesis	ADH (micelles)	Octane	Berezin, 1984
Synthesis of glycerides	Lipase (PEG-modif)	Benzene	Takahashi, 1984
Oil interesterification	Lipase (immob.)	Petroleum spirit	Wisdom, 1984
L-menthol production	Lipase (immob.)	Cyclohexane	Fukui, 1985

ACKNOWLEDGEMENTS

   To Professor Malcolm D. Lilly for his invaluable collaboration on this
work, most of which is taken from my Ph.D thesis, University College London,
1982. It is a pleasure to acknowledge the tireless assistance of
Miss Paula Morgado in the preparation of the manuscript.

REFERENCES

Aitio, A., 1977, Res. Commun. Chem. Pathol. Pharmac., 18:773.
Amis, E. and Hinton, J., 1973, "Solvent Effects on Chemical Phenomena",
     Vol. I., A.P.
Anusiem, A. and Oshodi, A., 1978, Arch. Biochem. Biophys., 189:392.
Balny, C. and Douzou, P., 1979, Biochimie, 61:445.
Bartling, G., Brown, H. and Chattopadhyay, S., 1974, Enzyme, 18:310.
Bayer, A.G., E.P. 22492, July 1979.
Bender, M., 1971, "Mechanisms of Homogeneous Catalysis from Protons to
     Proteins", Wiley-Interscience, New York.
Berezin, I.V. and Martinek, K., 1984, Ann. N.Y. Acad. Sci.
Brockman, H., Law, J. and Kezdy, F., 1973, J. Biol. Chem., 248:4965.
Buckland, B., 1974, Ph.D. thesis, University of London.
Buckland, B. Dunnill, P. and Lilly, M.D., 1975, Biotechnol. Bioeng.,
     17:815.
Butler, L., 1979, Enzyme Microb. Technol., 1:253.
Clement, G. and Bender, M., 1963, Biochemistry, 2:836.
Coetzee, J. and Ritchie, D., eds., 1969, "Solute-Solvent Interactions",
     Marcell Dekker, New York.
Connors, K. and Sun, S., 1971, J. Am. Chem. Soc., 93:7239.
Cordone, L. Izzo, V., Sgroi, G. and Fornili, S., 1979a, Biopolymers,
     18:1965.
Cordone, L., Cupane, A., San Biagio, P. and Vitrano, E., 1979b, Biopolymers,
     18:1975.
Coty, V., Gorring, R., Heilweil, I., Leavitt, R. and Srinivasan, S., 1971,
     Biotechnol. Bioeng., 13:825.
Covington, A. and Dikinson, T., eds., 1973, "Physical Chemistry of Organic
     Solvent Systems", Plenum Press, New York.
Cremonesi, P., Carrea, G., Ferrara, L. and Antonini, E., 1974, Eur. J.
     Biochem., 44:401.
Cremonesi, P., Carrea, G., Ferrara, L. and Antonini, E., 1975, Biotechnol.
     Bioeng., 17:1101.
Cremonesi, P., Mazzola, G. and Cremonesi, L., 1977, Annali Di Chimica,
     67:415.
Crosby, J. and Lienhard, G., 1970, J. Am. Chem. Soc., 92:5707.
Davies, J., 1954, Adv. Catalysis, 6:1.
Dawson, R., 1976, Biochim. Biophys. Acta., 452:413.
Dimroth, P., Guchhait, R., Stoll, E. and Lane, M., 1970, Proc. Nat. Acad.
     Sci. USA, 67:1353.
Dixon, M. and Webb, E., 1964, "Enzymes", 2nd ed., Longmans, London.
Douzou, P., 1977, Adv. Enz., 45:157.
Duarte, J. and Lilly, M.D., 1980, Enzyme Engineering, 5:363, Plenum Press.
Duarte, J., 1983, Ann. N.Y. Acad. Sci., Biochem. Eng. III, 413:548.
Edmonds, P., and Cooney, J., 1967, Appl. Microbiol., 15:411.
Eisenbach, M., Caplan, S. and Tanny, G., 1979, Biochim. Biophys. Acta.,
     554:269.
Epton, R., Hobson, M. and Marr, G., 1979, Enzyme Microb. Technol., 1:37.
Falcoz-Kelly, F., Baulieu E.-E. and Alfsen, A., 1968, Biochemistry, 7:4119.
Fukui, S. and Tanaka, A., 1985, Endeavour, 9:10.
Gibson, D., 1968, Science, 161:1093.
Gibson, G., Hensley, M., Yoshioka, H. and Mabry, T., 1970, Biochemistry,
     9:1626.

Hildebrand, J. and Scott, R., 1949, "The Solubility of non-Electrolyses", 3rd edition, Reinhold, New York.

Humphrey, A., 1967, Biotechnol. Bioeng., 11:3.

Humphrey, A., and Erickson, L., 1972, J. Appl. Chem. Biotechnol., 4:125.

Hymes, A., Robinson, D. and Canady, W., 1965, J. Biol. Chem., 240:134.

Ingalls, R., Squires, R. and Butler, L., 1975, Biotechnol. Bioeng., 17:1627.

Ingram, L., 1977, Appl. Environ. Microb., 33:1233.

Iio, M., Yamamoto, Y. and Ohta, N., 1978, Nippon Nôgeikagaku Kaishi, 52:493.

Ishida, H., Mori, M. and Tatibana, M., 1977, Arch. Biochem. Biophys., 182:258.

Iwasaki, 1976, Japan Kokai 7658901.

Johnson, R., 1978, Oxygenations with microorganisms, in: "Oxidation in Organic Chemistry", W. Trahanovsky, ed., Part C, Ch. 2., Academic Press, New York.

Jones, J. and Goodbrand, H., 1977, Can. J. Chem., 55:2685.

Jones, J. and Gordon, K., 1973, Biochemistry, 12:71.

Jones, J., Sih, C. and Perlam, D., 1976, Applications of biochemical systems in organic chemistry, in: "Techniques of Chemistry", A. Weissberger, ed., Vol. X (2 parts), John Wiley and Sons, New York.

Jones, J. and Schwartz, H, 1982, Can. J. Chem., 60:335.

Kelly, S., Butler, L. and Squires, R., 1976, Enzyme Technol. Dig., 5:107.

Kirschenbaum, D., 1963, Arch. Biochem., Biophys., 103:249.

Klibanov, A., Samokhin, G., Martinek, K. and Berezin, I., 1977, Biotechnol. Bioeng., 19:1351.

Klibanov, A., Sememov, A., Samokhin, G. and Martinek, K., 1978, Bioorganicheskaya Khimiya, 4:82.

Laidler, K., 1965, Faraday Soc. Discussions, 20:83.

Laidler, K. and Bunting, P., 1973, "The Chemical Kinetics of Enzyme Action", 2nd edition, Clarendon Press, Oxford.

Leavitt, R., Ryan, F. and Burgess, W., 1974, Adv. Exp. Med. Biol., 42:259.

Li, N., Brusca, D. and Mohan, R., 29 July 1979, US Patent 3,897,308.

Llor, J. and Cortijo, M., 1977, J. Chem. Soc. Perkin Trans., II:1111.

Lugaro, G., Carrea, G., Cremonesi, P., Casellato, M. and Antonini, E., 1973, Arch. Biochem. Biophys., 159:1.

Martinek, K., Levashov, A., Klyachko, N. and Berezin, I., 1977, Doklady Akademii Nauk SSSR, 236:920.

Maurel, P., 1978, J. Biol. Chem., 253:1677.

May, S. and Li, N., 1974, Enz. Eng., 2:77.

McGuiness, T., Brown, H., Chattopadhyay, S., and Chen, F., 1978, Biochim. Biophys. Acta., 530:247.

McLaren, A. and Packer, L., 1970, Adv. Enzymol., 33:245.

Melander, W. and Horvath, C., 1977, Arch. Biochem. Biophys., 183:200.

Melander, W. and Horvath, C., 1978, Enz. Eng., 4:355.

Misiorowsky, R. and Wells, M., 1974, Biochemistry, 13:4921.

Miura, Y., 1978, Adv. Biochem. Eng., 9:31.

Miles, J., Robinson, D. and Canady, W., 1963, J. Biol. Chem., 238:2932.

Mohammadzadeh-k, A., Feeney, R. and Smith, L., 1969, Biochim. Biophys. Acta., 194:241.

Myers, J. and Jakoby, W., 1973, Biochem. Biophys. Res. Commun., 51:631.

Nakahara, T., Erickson, L. and Gutierrez, J., 1977, Biotechnol. Bioeng., 19:9.

Nishikawa, A., 1978, Enz. Eng., 3:357.

Omata, T., Iwamoto, N., Kimura, T., Tanaka, A and Fukui, S., 1981, Eur. J. Appl. Microbiol., 11:199.

Oyama, K., Nishimura, S., Nonaka, Y., Kihara, K. and Hashimoto, T., 1981, J. Org. Chem., 46:5241.

Peterson, D. and Murray, H., 1952, J. Am. Chem. Soc., 74:1871.

Plishker, M., Thorpe, J. and Goldsmith, L., 1978, Arch. Biochem. Biophys., 191:49.

Poom, P. and Wells, M., 1974, Biochemistry, 13:4928.

Ralston, G., 1972, Compt. Rend. Trav. Lab. Carlsberg, 39:25.

Ray, A., 1971, Nature, 231:313.

Reichardt, C., 1979a, "Solvent Effects in Organic Chemistry", Verlag Chemie, Weinheim, New York.

Reichardt, C., 1979b, Angew. Chem. (Int. Ed.), 18.

Rohrschneider, L., 1973, Anal. Chem., 45:1241.

Royer, G. and Canady, W., 1968, Arch. Biochem. Biophys., 124:530.

Sanwall, B., Maeba, P. and Cook, R., 1966, J. Biol. Chem., 241:5177.

Schlusselberg, J. and Paredes, S., 1975, Biochim. Biophys. Acta, 405:89.

Schwartz, R. and Leathen, W., 1976, Petroleum Microbiology, in: "Industrial Microbiology", B. Miller and W. Litsky, eds., Ch. 13, McGraw-Hill, New York.

Schwartz, R. and McCoy, C., 1977, Appl. Environ. Microb., 34:47.

Sémériva, M., Chapus, C., Bovier-Lapierre, C. and Desnuelle, P., 1974, Biochem. Biophys. Res. Commun., 58:808.

Sémériva, M. and Desnuelle, P., 1976, Horizons Biochem. Biophys., 2:32.

Sémériva, M. and Desnuelle, P., 1979, Adv. Enzymol., 48:319.

Siegel, S. and Roberts, K., 1968, Space Life Sci., 1:131.

Snyder, L., 1974, J. Chromatogr., 92:223.

Snyder, L., 1978, J. Chromatogr. Sci., 16:223.

Svendsen, I., 1971, Compt. Rend. Trav. Lab. Carlsberg., 38:385.

Takahashi, K., Nishimura, H., Yoshimoto, T., Okada, M., Ajima, A., Matsushima, A., Tamaura, Y., Saito, Y., Inada, Y., 1984, Biotechnol. Lett., 6:765.

Tamm, C., 1974, FEBS Letters, 48:7.

Tan, K. and Lovrien, R., 1972, J. Biol. Chem., 247:3278.

Tang, J., 1965, J. Biol. Chem., 240:3810.

Torma, A. and Itzkovitch, I., 1976, Appl. Environ. Microb., 34:102.

Travers, F. and Hillaire, D., 1979, Eur. J. Biochem., 98:193.

Uzuki, T., Takashi, M. Noda, M., Komachiya, Y. and Wakamatsu, H., 23 September 1975, US Patent 3,907,638.

Verger, R., Mieras, M. and de Haas, G., 1973, J. Biol. Chem., 248:4023.

Walsh, C., 1979, "Enzymatic Reaction Mechanisms", W.H. Freeman and Co., San Francisco.

Wan, H. and Horvath, C., 1975, Biochim. Biophys. Acta, 410:135.

Weetall, H. and Vann, W., 1976, Biotechnol. Bioeng., 18:105.

Williams, P. and Worsey, M., 1976, J. Bacteriol., 125:818.

Wisdom, R., Dunnill, P., Lilly, M.D., Macrae, A., 1984, Enzyme Microb. Technol., 6:443.

Wishnia, A. and Pinder Jr., T., 1964, Biochemistry, 3:1377.

Wolf, R. and Luisi, P., 1979, Biochem. Biophys. Res. Commun., 89:209.

Yamada, M. and Mitsugi, K., 22 May 1978, Ajinomoto Co. Inc., Jap. Patent 78 56,385.

Yokozeki, K., Yamanaka, S., Takinami, K., Hirose, Y., Tanaka, A., Sonomoto, S. and Fujui, S., 1982, Appl. Microbiol. Biotechnol., 14:1.

Zahler, P. and Niggli, V., 1977, The use of organic solvents in membrane research, in: "Methods in Membrane Biology", E. Korn, ed., Vol. 8, Ch. 1, Plenum Press, New York.

ENZYMATIC ORGANIC SYSTEMS

Charles J. Sih, Shih-Hsiung Wu, and Yoshinori Kujimoto

School of Pharmacy, University of Wisconsin
Madison, Wisconsin 53706, USA

INTRODUCTION

Although many drugs currently in use are administered as racemates, there is reason to believe that the therapeutic effects of some racemic drugs reside primarily in one stereoisomer while its enantiomer may contribute to toxicity and adverse side effects (Simonyi, 1984). For example, the beta adrenergic blocking agent, propranolol [1-isopropulamino-3-(1-naphthyloxy)-2-propanol] is widely used for the treatment of angina pectoris, cardiac dysrhythmia, and hypertension. The commercial preparation (Inderal) is a racemic mixture of which only the S-(-)-enantiomer has beta adrenergic blocking activity (Howe and Shanks, 1966). The general notion that an enantiomerically pure pharmaceutical would be a better drug than the racemate demands the development of improved strategies for the synthesis of chiral compounds.

In recent years, interest in the use of enzymes as chiral catalysts has risen rapidly. Hydrolytic enzymes such as α-chymotrypsin, porcine liver esterase (PLE), lipases and other hydrolases from microorganisms are particularly suited for synthetic applications. These enzymes possess broad substrate specificity and many of them are highly enantioselective. Moreover, they have no coenzyme requirement and experimentally are easy reactions to carry out. Hence, a variety of bifunctional chirons have been prepared either _via_ enzymatic resolution of enantiomers or enantiotopically-selective hydrolysis of _meso_ and prochiral diesters (Huang et al., 1975; Chen et al., 1981; Gopalan and Sih, 1984).

On the other hand, enzymes of high stereochemical specificity are frequently not accessible and one has to resort to the tedious task of randomly screening for a suitable enzyme system. To circumvent this obstacle, we herein introduce a general strategy that extends the usefulness of enzymes of low to moderate enantioselectivity, so that they may now be effectively used for organic chiral syntheses. To facilitate the treatment of this concept, we shall first briefly discuss the general mechanistic principles, asymmetric catalysis and the kinetic resolution of enantiomers.

GENERAL MECHANISTIC PRINCIPLE

Asymmetric Catalysis

An esterase interacts differently with the enantiotopic features of a prochiral or meso substrate to form two diastereomers, ENZ-S and ENZ-S' as shown.

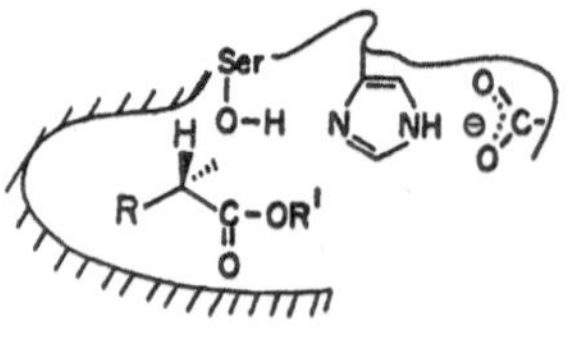

ENZ-S 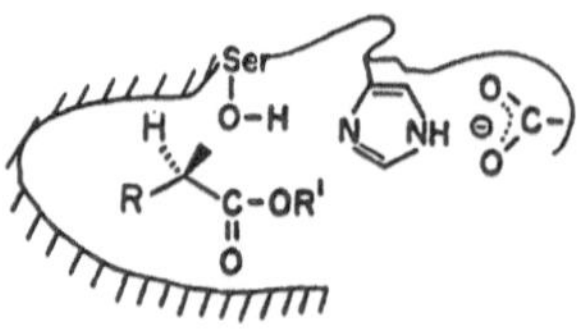   ENZ-S'

To achieve high enantioselection, two energetically different diastereomeric transition states must be formed during the course of the reaction. The enantiomeric excess (ee) is determined by the difference in free energy ($\Delta\Delta G^{\ddagger}$) of the transition states. A value of ($\Delta\Delta G^{\ddagger}$) of approximately 3 Kcal mole^{-1} would result in an essentially optically pure product.

A simplified enzymatic mechanism of ester hydrolysis is depicted in Scheme 1.

Scheme 1

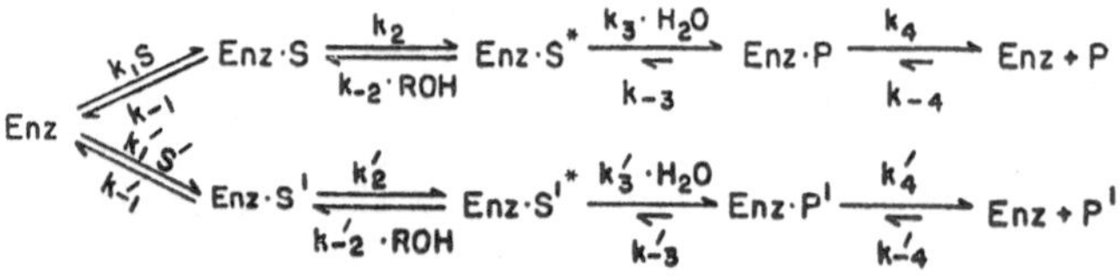

The first irreversible step in each pathway leading to the formation of P and P' enantiomers is the chirality determining step. In Scheme 1, when the concentration of ROH is low, the conversion of ENZ-S $\longrightarrow$ ENZ-S'* and ENZ-S' $\longrightarrow$ ENZ-S'* are virtually irreversible. Hence, it is the relative rates of these two parallel steps (k_2[ENZ-S] and k_2'[ENZ-S']) that determine the enantioselectivity (E'). For an enzyme system to exhibit maximal enantioselection, a rapid equilibration of the diastereomeric complexes

$$(ENZ\text{-}S \xrightleftharpoons[k_{-1}]{k_1} \quad E + S \xrightleftharpoons[k_{-1}']{} \quad ENZ\text{-}S')$$

must occur (i.e., $k_{-1} \gg k_1$ and $k_{-1}' \gg k_2'$), so that enantioselection is governed by equation 1.

$$E' = \frac{P}{P'} = \frac{[ENZ\text{-}S]k_2}{[ENZ\text{-}S']k_2} = \frac{k_2(k_1/k_{-1})}{k_2'(k_1'/k_{-1}')} \tag{1}$$

On the other hand, when the enzyme-complexes dissociation steps are slow (i.e., $k_{-1} \ll k_2$ and $k_{-1}' \ll k_2'$), enantioselection is now dictated by the relative rates of formation of ENZ-S and ENZ-S' (equation 2).

$$E' = \frac{P}{P'} = \frac{k_1}{k_1'} \tag{2}$$

In kinetic isotope effect studies, it is generally accepted that $k_1 = k_1'$. Although the value of k_1 is probably not too much different from that of k_1' in enantioselective processes, this difference may be significant. For example, in the asymmetric hydrogenation of methyl-z-(α)-

acetamido cinnamate, catalyzed by $[Rh(\underline{R}, \underline{R}\text{-DiPAMP})]^+$, the ratio of k_1/k_1' was determined to be 2, corresponding to approximately 30% $\underline{ee}$ (Howe and Shanks, 1966). The reported observed values (Hammes, 1982) of k_1 for enzymatic reactions range from 10^6 to 10^6 mole $\sec^{-1}$ reflecting that k_1 is not purely a diffusion constant but includes binding and conformational effects. In any event, low enantioselection results if the enzyme-complex dissociation steps (i.e., k_{-1} and k_{-1}') are slow or if the substrate binds too tightly to the enzyme.

Kinetic Resolution of Enantiomers

The enzyme binds with each chiral enantiomer to form two disastereomeric complexes, ENZ-S and ENZ-S'. The principles governing enantioselection (equations 1 and 2) are the same as before.

Although numerous kinetic resolution experiments have been conducted using hydrolytic enzymes, with few exceptions, these data have not been quantitatively analyzed to allow comparison of enantioselectivity. Consider a typical kinetic resolution experiment using two different ester hydrolases, where one of the enantiomers is preferentially hydrolyzed as shown:

For enzyme 1, the reaction is terminated at 30% conversion (C = 0.30) and the optical purity, expressed as enantiomeric excess ($\underline{ee}$) of the product fraction ($\underline{ee}_P$) and the remaining substrate ($\underline{ee}_S$), was determined to be $\underline{ee}_{(S)}$ = 0.38 and $\underline{ee}_{(P)}$ = 0.89. For enzyme 2, the reaction is terminated at 60% conversion ($\overline{C}$ = 0.60) with $\underline{ee}_{(S)}$ = 0.90 and $\underline{ee}_{(P)}$ = 0.60. The question now arises: which of the two enzymes is the more enantioselective one? One obvious way to compare enanioselectivity would be to terminate both enzymic reactions at the same extent of conversion. However, experimentally this is tedious to achieve, because the kinetics of hydrolysis must be continuously monitored carefully. Even then, it would be difficult to terminate both reactions __exactly__ at the same extent of conversion. Consequently, there is a need to express enantioselectivity in terms of an index that is independent of the variable, C. The enantiomeric ratio (Chen et al., 1982), E, is a biochemical constant that is independent of time and substrate concentration; it is related to C and $\underline{ee}_{(S)}$ by equation 3.

$$\frac{\ln([1 - C][1 - ee_{(S)}])}{\ln([1 - C][1 + ee_{(S)}])} = E \tag{3}$$

Thus, substituting the values of $\underline{ee}_{(S)}$ and C for the two enzymes, the E values for enzymes 1 and 2 are calculated to be 24 and 12, respectively. Although the value of C is usually experimentally determined via methods such as GC or HPLC, yet it is much more accurate and convenient to calculate C from the experimentally-derived values of $\underline{ee}_{(S)}$ and $\underline{ee}_{(P)}$ using equation 4.

$$C = \frac{ee_{(S)}}{ee_{(S)} + ee_{(P)}} \qquad (4)$$

Hence, in kinetic resolution experiments, if the time independent values of $ee_{(S)}$ and $ee_{(P)}$ are defined, the values of E and C may be calculated accurately from equations 3 and 4, respectively.

Multiplicity of Enzymes in Commercial Preparations and Microorganisms

Most of the hydrolytic enzymes used for synthetic applications are either commercial preparations or intact microorganisms. In many instances, the hydrolytic reaction is only partially enantioselective, which may be interpreted in two ways. The enzyme system contains a single hydrolase which is only partially enantioselective. Alternatively, the preparation may contain more than one hydrolase generating products of opposite configuration at different rates (depending on V and K). These two conditions may be readily distinguished from each other if one conducts the reaction at several different substrate concentrations. In the former case, enantioselectivity is independent of changes in substrate concentration, whereas the latter is not (Chen et al., 1984). At low substrate (S) concentrations, (S——>0), the relative rates of the competing reactions depend on V/K (apparent first order rate constant) and the ee of the product is governed by equation 5; whereas at high substrate concentrations (S——>∞), the ee is dictated by equation 6, where V_R, V_S and K_R, K_S denote maximal velocities and Michaelis constants for enzymes R and S, respectively (Chen et al., 1984).

$$ee = \frac{\dfrac{V_R}{K_R} - \dfrac{V_S}{K_S}}{\dfrac{V_R}{K_R} + \dfrac{V_S}{K_S}} \qquad (5) \qquad\qquad ee = \frac{V_R - V_S}{V_R + V_S} \qquad (6)$$

Consequently, when competing enzymes of opposing chirality are present in the preparation, it is often possible to enhance the optical purity of the product by conducting the reaction at low substrate concentrations (i.e., continuous slow feeding), because the high V/K enzyme would be the major contributor in the competing reaction.

RESULTS AND DISCUSSION

Asymmetric Synthesis

Recycling of carboxy half-esters. It is generally believed that a substance is attached to the active site of the enzyme at a minimum of three different loci (Ogsten effect) to achieve enantioselection (Bentley, 1969) Hence, by altering substituents on the substrate molecule, the degree of enantioselection may be enhanced or diminished. For hydrolases, it is convenient to alter the ester group because chemically it is a trivial manipulation. Unfortunately, at the present time there is insufficient concrete data to allow the prediction of the effect of the size of the ester grouping vs. enzymic enantioselectivity.

When a prochiral or meso dicarboxylic ester (A) is exposed to carboxy-esterases, the reaction generally teminates at the monoester stage in most cases (Levy and Ocken, 1969). Thus, the ratio of the rates of formation of the R-enantiomer and the S-enantiomer is dictated by the constant $\alpha = k_1/k_2$ and the optical purity of the monester fraction is simply defined by $\beta = (\alpha - 1)/(\alpha + 1)$.

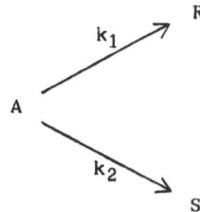

Pig liver esterase (PLE) is an enzyme widely used in asymmetric syntheses because of its ability to catalyze enantiotopic group differentiation (Huang et al, 1975; Chen et al., 1981; Gopalan and Sih, 1984). When methyl (1), ethyl (2) and n-propyl (3) diesters of cis-2,4-dimethylglutaric acid were exposed to PLE[1], the corresponding chiral half-esters (4-6) were obtained. The absolute configuration of the half-esters were established by their reduction with $LiBH_4$ to yield the lactone (2S,4R)-7 (Jakovac et al., 1980). Since $LiBH_4$ selectively reduces carboxylic acid esters and not carboxylic acids (Cornforth et al., 1966), it follows that PLE preferentially cleaved the pro-S ester grouping of these diesters (1-3). It is noteworthy that the enantioselectivity of hydrolysis improved as the size of the ester grouping is enlarged. The enantiomeric excesses (ee) were as follows: 6 (0.83); 5 (0.63); 4 (0.64).

H3C   CH3   PLE   H3C   CH3   LiBH4   H3C   CH3

RO_2C CO_2R ⟶ HO_2C CO_2R ⟶

1	R = CH_3		4	R = CH_3	
2	R = C_2H_5		5	R = C_2H_5	
3	R = n-C_3H_7		6	R = n-C_3H_7	

7

To improve the optical purity of 4 (ee = 0.64), it was converted into the mixed diester 8 via reaction with propyl iodide and tetramethylammonium hydroxide (Greeley, 1974). When 8 was incubated with PLE, the n-propyl ester group was again preferentially cleaved to give back 4 (57% yield, reaction terminated at 70% conversion). Analysis revealed that the optical purity was indeed enhanced to ee = 0.85.

H3C   CH3   PLE   ⟶   4

RO_2C CO_2CH_3 ee = 0.85

8   R = n-C_3H_7

ee = 0.64

---

[1] The incubation mixture contained: 200 mg each of 1 or 2 or 3, suspended in 20 ml or 0.1 M phosphate buffer, pH 8.0. The following amounts of PLE were used: for 1, 100 units; 2, 150 units; and 3, 600 units; incubation times were 60 min for 1; 70 min for 2; and 3 hours for 3.

This moderately successful observation encouraged us to examine further the generality of this recycling method. Thus, the methyl (9), ethyl (10), and n-propyl (11) or 3-methylglutaric acid were incubated with PLE.[1] The absolute configurations of the resulting half-esters (12-14) were established by their transformation into (-)(S)-3-methylvalerolactone (15) (Irwin and Jones, 1977), indicating that PLE again preferentially cleaved the pro-S ester grouping of these diesters. However, in this instance the enantioselectivity decreased as the size of the ester grouping was enlarged.

ee

9 R = CH_3	12 R = CH_3	0.84
10 R = C_2H_5	13 R = C_2H_5	0.70
11 R = n-C_3H_7	14 R = n-C_3H_7	0.40

15

The relative rates[2] of hydrolysis of 9, 10, and 11 by PLE were approximately 100:94:38, respectively. Hence, we surmised that the optical purity in 14 might perhaps be enriched by recycling, assuming that PLE would selectively cleave the methyl ester grouping in 16a. In essence, one utilizes a subsequent kinetic resolution step to enhance the optical purity of 14. However, contrary to our expectations, when 16a, prepared by diazomethane treatment of 6, was exposed to PLE, it was rapidly transformed into a 9:1 mixture of 17a and 14a, respectively, which were not separable by conventional silica gel chromatography.

16a R = n-C_3H_7

14a R_1 = H; R_2 = n-C_3H_7

17a R_1 = CH_3; R_2 = H

Furthermore, the enantiomer 16b, prepared by reaction of 12 with propyl iodide and tetramethylammonium hydroxide, was converted by PLE to a 98:2 mixture of 17b and 14b respectively.

---

[1]The reaction mixture contained 200 mg each of the respective substrate, suspended in 20 ml of 0.1 M phosphate buffer, pH 8.0. For 9 and 10, 100 units of PLE (Sigma) were used and the contents were incubated for 50 min. For 11, 400 units of enzyme were used and the incubation time was extended to 180 min. The half esters were isolated by acidification of the reaction mixture to pH 2.0 and extraction with ethyl acetate. After the usual work-up, the crude products were purified with SiO_2 column chromatography (elution with hexane-ethyl acetate, 4:1).

[2]The time required to hydrolyze 50% of 9, 10, and 11 was estimated by GLC [1.5% OV-101 on Chromsorb GJP column (5 ft)] analyses of the amount of residual substrate at various time intervals; column temperatures of 120, 140 and 160°C were used for 9, 10 and 11, respectively.

H3C\\\\H
RO2C   CO2CH3
      ⟶
H3C\\\\H
R1O2C   CO2R2

16b R = n-C$_3$H$_7$

14b R$_1$ = n-C$_3$H$_7$; R$_2$ = H

17b R$_1$ = H; R$_2$ = CH$_3$

These results demonstrate that no enantiomeric enrichment was achieved by recycling, for PLE preferentially cleaved the propyl ester grouping in both enantiomers 16a and 16b. Because this recycling method does not appear to have general applicability, we turned our attention to an alternative approach.

OPTICAL PURITY ENHANCEMENT OF MONOACETOXY ESTERS

S is an achiral diacetoxy ester with a plane of symmetry, and it is converted by a hydrolase into the two enantiomeric monoesters, P (fast-forming) and Q (slow-forming); these are further hydrolyzed by the same enzyme to yield the diol, R.

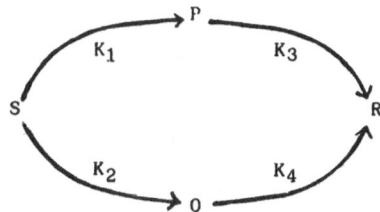

If the enzyme has the same stereochemical preference in the second step so that $K_1 > K_2$ and $K_4 > K_3$, the combined procedures of enantioselective hydrolysis and concomitant kinetic resolution provide a convenient method of enriching the optical purity of the monoester fraction (P + Q). Equations have been derived to calculate the three kinetic parameters, E_1, E_2 and α (Wang et al., 1984). These constants allow the prediction of the enantiomeric excess (ee) of the monester fraction and the optimization of optical and chemical yields.

To test the validity of this theory, the meso-diester, 1,5-diacetoxy-cis-2,4-dimethylpentane (18) was exposed to PLE. At various intervals, the amount of 18, 19, and 20 and the ee of the monoacetate fraction (19) were quantitatively determined.[1] From these results, the kinetic constants for the hydrolysis of 18 were calculated to be: α = 2.47 ± 0.36; E_1 = 0.22 ± 0.05, and E_2 = 0.60 ± 0.10 (mean ± S.D.). It is noteworthy that even with α as low as 2.47, it is still possible to prepare the monoacetate (19) of high optical purity (ee = 0.80, 36% yield; ee = 0.95, 15% yield). These kinetic constants allow one to predict the ee of the monoester fraction for any degree of conversion (Fig. 1). To establish the absolute stereochemistry of 19, a sample of 19 was transformed into (2R, 4S)-2,4-dimethylvalerolactone (Chen et al., 1981) (21) by Jones oxidation (CrO$_3$/H$_2$SO$_4$), ester hydrolysis (1 N NaOH) and conversion to the lactone

---

[1]The amount of 18, 19, and 20 was quantitatively measured by GLC analysis (5% OV-101 on Chromosorb WHP column, 3 ft); the temperature of the column was 100°C. The enantiomeric excess (ee) of the monoacetate fraction (19) was determined by ^{1}H NMR spectroscopy (CCl$_4$) in the presence of Eu(hfc)$_3$.

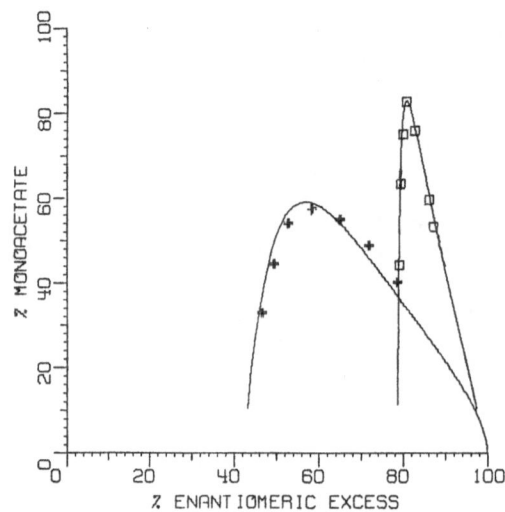

Fig.1 Percent monoacetate as a function of
percent enantiomeric excess (ee).
+-+: 18 with PLE; □-□: 26
with PLE. The curves are computer
generated using the values of α, E_1
and E_2 given in the text. □ and
+ are experimentally determined
values.

(TsOH). This result confirms that PLE preferentially cleaved the pro-R-
acetoxy group of 18.

18   19 R = CH_3CO 21

20 R = H

Optically-active 4-oxo-2-cyclopentenyl acetates are valuable
intermediates for the stereoselective and enantioselective synthesis of
cyclopentanoid natural products (Hane et al., 1982). Because 4(R)-hydroxy-
cyclopent-2-enone (22) is an important starting material for the synthesis
of prostaglandins (Bindra and Bindra, 1977), it is an attractive target for
enzymatic organic synthesis. One of the shortest and simplest routes to
prostagladins involves the 1,4 -addition of the β-chain to 4(R)-alkoxy-
cyclopent-2-enone (23) and the trapping of the resulting enolate (24) with
the α-chain (Stork and Isobe, 1975). In the conjugate addition, the β-chain
approaches trans to the alkoxy group and the substituent trapping of the
enolate gives the desired (all trans orientation) absolute configuration as
in natural prostaglandins (Sih et al., 1975). This process has now been
optimized and is commonly known as the three-component coupling process
(Suzuki et al., 1983).

22 R = H

23 R = THP

4($\underline{R}$)-Hydroxycyclopent-2-enone ($\underline{22}$) had been prepared in very low overall yield via a combination of Bakers' yeast hydrolysis in 3,5-diacetoxycyclo-pent-1-en and chemical oxidation (Tanaka et al., 1976). The need for an improved process for the preparation of $\underline{22}$ prompted us to examine our strategy of using the $\underline{meso}$-diacetoxy ester, $\underline{26}$, as the substrate. The diacetate $\underline{26}$ was exposed to PLE at pH 7.0 and at various intervals the extent of conversion ($\underline{26}$, $\underline{27}$ and $\underline{28}$) were determined.[1] These data were used to calculate the three kinetic constants (Bindra and Bindra, 1977): α = 8.44 ± 0.56, E_1 = 0.06 ± 0.01, and E_2 = 0.12 ± 0.02, which reveals that the maximal recovery of the monoacetate, $\underline{27}$, was 83% with an $\underline{ee}$ of 0.81 (Fig.1). Since $\underline{27}$ is a crystalline solid, it can be readily crystallized from benzene-Skelly B (1:5) to yield $\underline{27}$ with an $\underline{ee}$ = 0.96. The absolute configuration of $\underline{27}$ was established by its transformation into the known (+)-3($\underline{S}$)-hydroxy-5($\underline{R}$)-(tetrahydropyranoxy)-cyclopent-1-ene (Nara et al., 1980), confirming that the $\underline{pro}$-$\underline{R}$ acetoxy group of $\underline{26}$ was preferentially attacked by the PLE.

The essence of this approach lies in the recognition of the inherent consecutive kinetic resolution step in enhancing the optical purity of the monoacetate fraction during enantioselective hydrolysis of diesters. The successful preparation of the chirons $\underline{19}$ and $\underline{27}$ of high optical purities illustrates the general applicability of this strategy to biochemical processes involving enantiotopic group differentiation. Further, this method extends the usefulness of hydrolases of low to moderate enantioselectivity, so that synthetic chemists now have more flexibility in their selection of enzyme systems for synthetic applications.

Kinetic Resolution of Enantiomers

In analyzing the enzymatic kinetic resolution of the racemic $\underline{trans}$-diacetates, $\underline{29a}$ (3$\underline{R}$, 5$\underline{R}$) and $\underline{30a}$ (3$\underline{S}$,5$\underline{S}$), the stereochemical outcome is dependent on the relative rates of the four hydrolytic steps, where k_1, k_2, k_3, k_4 (the apparent first order rate constants) are related to their corresponding kinetic constants of the enzyme, V_1/K_1, V_2/K_2, ... (V = maximal velocity; K = Michaelis constant).

---

[1]The amounts of $\underline{26}$, $\underline{27}$ and $\underline{28}$ were quantitatively measured by GLC analysis (5% OV-101 on Chromosorb WHP column, 3 ft at 95°C). The enantiomeric excess ($\underline{ee}$) of the monoacetate fraction ($\underline{27}$) was determined by ^{1}H NMR spectroscopy ($\overline{CCl_4}$) in the presence of Eu(hfu)$_3$.

AcO·····⟨ ⟩·····OAc $\xrightarrow{k_1}$ AcO·····⟨ ⟩·····OH $\xrightarrow{k_2}$ HO·····⟨ ⟩·····OH

  29a   29b   29c

AcO·····⟨ ⟩·····OAc $\xrightarrow{k_3}$ AcO·····⟨ ⟩·····OH $\xrightarrow{k_4}$ HO·····⟨ ⟩·····OH

  30a   30b   30c

In principle, one would expect the enzyme to have the same stereochemical preferences for the diacetate (29a + 30a) and the monoacetate (29b + 30b). That is, when $k_1 > k_3$; $k_2 > k_4$. Hence, when an enzyme has a low to moderate enantioselectivity for the first hydrolytic step, it should be possible to enrich the optical purity of the monoacetate (29b + 30b) by taking advantage of the inherent successive hydrolytic step. However, the relative rates of the two hydrolytic steps must be of the same order of magnitude. For example, if an enzyme preferentially cleaves the pro-R acetate, the monoacetate would be enriched in 30b and the diol would be enriched in 29c.

When the racemic trans-diacetate (29a + 30a) (220 mg) was exposed to PLE (80 units) in 15 ml of phosphate buffer at pH 7, it was rapidly and completely transformed into the monoacetate (29b + 30b) before the appearance of any diol (29c + 30c). To determine the enantioselectivity of the first hydrolytic step, the reaction was terminated after 3 hours and the residual diacetate [29a + 30a; $[\alpha]_D$ -55° (MeOH) (Miura et al., 1976); 29a + 231° (MeOH)] and the monoacetate [29b + 30b; $[\alpha]_D$ + 52° (MeOH) (Miura et al., 1976); 29b + 255° (MeOH)] were isolated. These data allowed us to determine the values of $ee_{(S)}$ = 0.24 and $ee_{(P)}$ = 0.20. In turn, the extent of conversion, C was calculated [equation (4)] and an E value of 2.1 ± 0.4 was obtained from equation (3). This result demonstrates that PLE preferentially cleaved the pro-R acetoxy of the trans-diacetate (29a + 30a).

Prolonged exposure (7 hrs) of the trans-diacetate (29a + 30a) to PLE resulted in the formation of the monoacetate (29b + 30b) and the diol (29c + 30c). The optical rotation of the residual monoacetate (29b + 30b) was $[\alpha]_D$ -209° (MeOH) whereas the diol (29c + 30c) was $[\alpha]_D$ + 75° (MeOH) (reported Miura et al., 1976); 30c $[\alpha]_D$ -237° (MeOH). These results reveal that PLE maintained the same stereochemical preference in the second hydrolytic step in preferentially cleaving the pro-R ester of 29b. However, the enantiomeric ratio, E, was only 4.4. It is noteworthy that PLE consistently preferentially cleaved the pro-R acetoxy ester. Although the stereochemical behaviour is in accord with our postulate, the large variance in the rates of the two hydrolytic steps violates one of the requisites of this enantiomeric enrichment strategy. Perhaps by simply changing the ester grouping of the substrate, one may be able to reduce this difference in the rates of the two steps. Nevertheless, it is our contention that this concept is useful, for it may be applicable to the kinetic resolution of many other trans-diacetates.

ACKNOWLEDGEMENTS

This investigation was supported in part by Grant GM33149 of the National Institute of Health.

REFERENCES

Bentley, R., 1969, "Molecular Asymmetry in Biology", Vol. I., Academic Press, New York, p 148.

Bindra, J.S. and Bindra, R., 1977, "Prostagladin Synthesis", Academic Press, New York, and references cited therein.

Chen, C.S., Fujimoto, Y. and Sih, C.J., 1981, J. Am. Chem. Soc., 103:1380.

Chen, C.S., Fujimoto, Y., Girdaukas, G. and Sih, C.J., 1982, J. Am. Chem. Soc., 104:7294.

Chen, C.A., Zhou, B.N., Girdaukas, G., Shieh, W.R., van Middlesworth, F., Gopalan, A.S. and Sih, C.J., 1984, Bioorganic Chem., 12:98.

Cornforth, J.W., Cornforth, R.H., Popjak, G. and Yengogan, L.S., 1966, J. Biol. Chem., 241:3970.

Gopalan, A.S. and Sih, C.J., 1984, Tetrahedron Lett., 5235.

Greeley, R.H., 1974, J. Chromatography, 88:229.

Halpern, J., 1982, Science, 217:401.

Hammes, G.G., 1982, "Enzyme Catalysis and Regulation", Academic Press, New York, p 101.

Hane, M., Raddatz, P., Walenta, R. and Winterfeldt, E., 1982, Angew, Chem. Int. Ed. Engl., 21:480.

Howe, R. and Shanks, R.G., 1966, Nature, 210:1336.

Huang, F.C., Lee, L.F.H., Mittal, R.S.D., Ravikumar, P.R., Chan, J.A., Sih, C.J., Caspi, E. and Eck, C.R., 1975, J. Am. Chem. Soc., 97:4144.

Irwin, A.J. and Jones, J.B., 1977, J. Am. Chem. Soc., 99:556.

Jakovac, I.J., Ng, G., Lok, K.P. and Jones, J.B., 1980, J. Chem. Soc. Chem. Commun., 515.

Levy, M. and Ocken, P., 1969, Arch. Biochem. Biophys., 135:259.

Miura, S., Kurozumi, S., Toru, T., Tanaka, T., Kobayashi, M., Matsubara, S. and Ishimoto, S., 1976, Tetrahedron, 32:1893.

Nara, M., Terashima, S. and Yamada, S., 1980, Tetrahedron, 36:3161.

Sih, C.J., Heather, J.B., Sood, R., Price, P., Peruzzotti, G., Hsu Lee, L.F., and Lee, S.S., 1975, J. Am. Chem. Soc., 97:865.

Simonyi, M., 1984, Med. Chem. Rev., 4:359.

Stork, G. and Isobe, M., 1975, J. Am. Chem. Soc., 97:6260.

Suzuki, M., Yanagisawa, A.and Noyori, R., 1983, Tetrahedron Lett., 24:1187.

Tanaka, T., Kurozumi, S., Toru, T., Miura, S., Kobayasi, M. and Ishimoto, S., 1976, Tetrahedron, 32:1713.

Wang, Y.F., Chen, C.S., Girdaukas, G. and Sih, C.J., 1984, J. Am. Chem. Soc. 106:3695.

HARNESSING BIOLOGICALLY-CATALYSED ELECTRON TRANSFER REACTIONS FOR BIOSENSORS

I.J. Higgins, M.F. Cardosi and A.P.F. Turner

Biotechnology Centre, Cranfield Institute of Technology
Cranfield, Bedford MK43 OAL, England

SUMMARY

There is a need in many areas of human activity for simple, rapid, reliable, in situ chemical analysis. One approach to achieving such an analytical revolution involves the exploitation of biological systems in association with electronic technology. This contribution briefly describes some recent work in this laboratory aimed at developing amperometric devices.

INTRODUCTION

We live in an environment which contains a host of substances that can influence, aggravate or stimulate various aspects of our health and behaviour. It is not surprising therefore, that in the last decade there has been a tremendous interest in developing a range of detection technologies designed to monitor these variables. Recently, this development has been accelerated by the realisation that detection systems, particularly those based on biological processes, could prove to be of commercial significance. At the same time, it was quickly appreciated that these devices (biosensors) could revolutionize several areas of analytical chemistry including health care, veterinary medicine and pollution monitoring.

Undoubtedly, the major impetus for the development of biosensors has come from the advances in health care technology. For example, it is now generally accepted that the frequent measurement of biochemical parameters such as plasma electrolytes, gases and metabolites is an essential prerequisite for the diagnosis of possible maladies and the subsequent administration of patient care. Thus, cheap and reliable in vitro sensors are required for monitoring key metabolites which can be used both in the hospital ward and at home.

The continuous signal provided by some categories of biosensor may be exploited for the continuous in vivo monitoring of metabolically unstable patients. Such in vivo sensors could monitor key metabolites either intravascularly or transcutaneously and provide continuous information on the metabolic changes occurring in their immediate environment. Continuous monitoring by such implantable sensors could in theory provide sufficient real time data to direct drug release via an associated mechanical device. Thus, they could be incorporated into biofeedback systems such as an artificial pancreas and thereby alleviate the need for self monitoring and insulin injection regimes in diabetic patients (Turner and Pickup, 1985).

In terms of industrial applications it is envisaged that biosensors will play an important role in fermentation control. Sensors will be developed which will monitor process variables such as substrate concentration, biomass, dissolved gases, ions, carbon sources and so on. As in the above example, such on line biosensors would provide sufficient information for the feed back control of the fermentation process thus ensuring that the right conditions are maintained for the maximum yield of the desired product.

THE BIOSENSOR

A biosensor is a device which converts a biological recognition process into an electrical signal the amplitude of which is related to the concentration of analyte in the solution. In general terms, the biosensor consists of a biological component such as an enzyme, antibody, organelle, whole cell or tissue held in close proximity to a suitable transducer. In this configuration the biological component plays a dual role. First, it confers selectivity upon the device and secondly it produces or consumes a species which can be detected by the transducer.

The intimate contact between the biological component and the transducer facilitates both a rapid response and high sensitivity. The immoblization of the biocatalyst to the transducer can be achieved in a number of ways:

(1) By covalently linking the biocatalyst to an insoluble matrix via a bidentate ligand.

(2) By occluding the biocatalyst within a polymer matrix such as polyacrylamide or agarose.

(3) By trapping the biocatalyst behind a semipermeable membrane such as nylon, cellophane or cellulose acetate.

(4) By covalently attaching the biocatalyst directly to the surface of the transducer. In this configuration not only is an intimate contact established but also diffusional constraints observed with polymer and membrane systems can sometimes be eliminated.

The transducer in a biosensor device is an electrical device which responds to the changing physico/chemical parameters associated with the specific interaction of the biocatalyst and its substrate. Although the configuration of the transducer will ultimately depend on the nature of species produced or consumed by the biocomponent, as a rule the design of the transducer should accomodate the following desirable features:

(1) It must be highly specific for the analyte of interest and respond in the appropriate concentration stage.

(2) The response should be fast and the signal produced should be amenable to some form of manipulation such as amplification, storage and display.

(3) The transducer should be amenable to miniaturisation.

The transducers used in biosensor applications may be conveniently categorised into four main types:

(1) Electrochemical
(2) Optical
(3) Thermometric
(4) Other

ELECTROCHEMICAL SENSORS

(a) Potentiometric Sensors

In a potentiometric based sensor an ion specific membrane or sensing surface acts like a battery generating a potential difference which is proportional to the logarithm of analyte concentration (activity) in the solution. The potential obtained with such a device is measured relative to an inert reference electrode such as the calomel electrode or the silver/ silver chloride electrode. Because the membranes used in these probes tend to be poor conductors, the measurements are made under conditions of essentially no current flow (equilibrium). Perhaps the best known examples of potentiometric sensors are the ion selective electrodes (ISEs). Examples of these include the pH probe, pCO_2 and O_2 monitors and the K^+ electrode.

The basis of an ion selective electrode is a membrane that will selectively extract one type of charged species from the solution into the membrane phase. This is shown schematically in Figure 1. The resulting charge difference generated across the sample-membrane interface will then be released to the concentration of target ions in the solution. The membrane in this context may be either glass, as in the case of the pH probe and Na^+ ISE, a solid crystal, ion exchange material or PVC dispersed with a number of selective binding sites.

The best example of a potentiometric biosensor is the potentiometric enzyme electrode. Here the membrane is a multilayer composite containing one or more immobilized enzymes. When placed in contact with a suitable substrate, the analyte, the membrane catalyses the conversion of the substrate into an electroactive species. Thus for example urease, an enzyme which catalyses the hydrolysis of urea as shown below, may be exploited for the determination of urea by immobilization around a pH, NH_4^+, HCO_3^- or CO_2 electrode. (Guilbault and Shu, 1971).

Guilbault and Shu extended the idea of their urea determining electrode to monitor tyrosine by associating decarboxylase activity with a pCO_2 electrode (Guilbault and Shu, 1972). Similarly, probes for phenylalanine and lysine have also been reported using the appropriate amino acid decarboxylase (Calvot et al., 1975). In 1973 Nagy et al. described an enzyme electrode for glucose based on the following sequence of reactions.

$$(1) \quad glucose + O_2 + H_2O = gluconic\ acid + H_2O_2$$
$$(2) \quad H_2O_2 + 2I^- + 2H^+ = I_2 + H_2O.$$

Horse radish peroxidase was used in stage (2) to give high efficiency at low concentrations (10^{-4}M) of iodide. The enzyme catalyses the oxidation of I^- by the peroxide produced from the oxidation of glucose. This results in a depletion of I^- which can be monitored with an I^- ISE.

A similar method for glucose detection was reported by Siddiqi (Siddiqi, 1982). The technique used an organofluorine compound which in the presence of horse radish peroxidase, reacted with the peroxide generated by the action of glucose oxidase to give free fluoride ions. These ions were then detected with a F^- ISE.

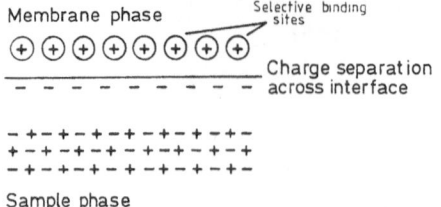

Fig. 1. Operational features of an
ion selective electrode

The use of penicillinase (β-lactamase) in conjunction with a pH
electrode gives a convenient sensor for the penicillins. The operational
principle for such devices relies on detecting the decrease in pH which
accompanies the hydrolysis of the β-lactam. (Nilsson et al., 1973;
Papariello et al., 1973).

(b) Amperometric Sensors

Another class of electrochemical device is the amperometric sensor.
Here a voltage is applied between a working and a reference electrode with
the result that the imposed potential encourages electron transfer (redox)
reactions to take place causing a net current to flow. The magnitude of the
current is proportional to the concentration of electroactive material
present in the solution.

Perhaps the best known example of a conventional amperometric sensor is
the Clark pO_2 electrode. In this system oxygen diffuses through a gas
permeable teflon membrane and is reduced to hydrogen peroxide at a bare
platinum electrode kept at a fixed potential (-600 mV) vs the silver-silver
chloride reference.

This system has proved useful for the development of amperometric
biosensors because of the wide variety of naturally occurring enzymes
(oxidases and oxygenases) which carry out their function with the net
consumption of oxygen. Such enzymes catalyse the oxidation of a wide
variety of compounds such as hydrocarbons, steroids, fatty acids, hydroxy
acids, amino acids, purines, pyrimidines, aldehydes, carbohydrates and
phenols.

Initial experiments of this type were carried out using glucose oxidase
entrapped in polyacrylamide gels held in close proximity to the surface of
the oxygen probe (Updike and Hicks, 1967). The system is based on the
reaction:

$$glucose + O_2 + H_2O = gluconolactone + H_2O_2.$$

When the electrode is placed in contact with a solution containing both
glucose and oxygen the substrates diffuse into the immobilised layer and
react with the enzyme. In the presence of glucose the rate of diffusion of
oxygen to the surface of the platinum is reduced resulting in a decreased
steady state current. A linear relationship is observed between current and
glucose concentration as long as the concentration of glucose is below the
Km(app) for the immoblized enzyme and oxygen is present in non rate limiting
excess.

An extension of this technology has led to the development of a family
of sensors for the detection of L-amino acids based on L-amino acid oxidase

(Nanjo and Guilbault, 1975). Oxidation of L-amino acids by this enzyme takes place according to the overall equation:

$$\underset{\substack{|\\ NH_2}}{R\text{-}C\text{-}COOH} + O_2 + H_2O = \underset{\substack{\|\\ O}}{R\text{-}C\text{-}COOH} + NH_3 + H_2O_2$$

Although the aminoacid oxidase used showed stereospecificity it was not very selective for the side chain with the result that the probe responded to a number of amino acids, albeit with different sensitivities.

Glucose oxidising bacteria have also been used in place of purified enzyme. Karube et al. (1979) described a system using immobilised Pseudomonas fluorescens and an oxygen electrode for fermentation monitoring.

Although extremely flexible, there are two inherent disadvantages with systems that rely on the polarographic determination of oxygen. First there is the high potential needed to bring about the reduction of oxygen at the naked platinum surface. This can introduce the problem of interference from other species in the solution which are electroactive within the potential window. Secondly there is the susceptibility of the technique to changes in the dissolved oxygen tension of the sample. It follows from this that such devices could not be used under anaerobic conditions though improvements can be made by utilising membranes which trap a layer of molecular oxygen close to the surface of the probe (Clark and Duggan, 1982).

One solution to the above problems is to exploit alternative electron sinks, mediator molecules, which can interact with the reduced enzyme and then be reoxidised at the working electrode. For example, Williams et al. (1970) described a system based on glucose oxidase and benzoquinone as the electron acceptor. The mechanism for this can be summarized as:

By controlling the amount of artificial electron acceptor present measurements could be made which were independent of local O_2 tension. Furthermore, because the sensor now operated at the oxidation potential of the quinone (400mV vs SCE) the problem of interference was reduced. Several artificial electron acceptors for glucose oxidase have been proposed including 2,6,dichlorophenolindophenol, methyleneblue and ferricyanide.

A mediator should fulfil the following criteria for biosensor application:

(1) It should be easily reduced by the enzyme.

(2) It should be rapidly reoxidized at the working electrode.

(3) The overpotential for the oxidation should be low and pH independent.

(4) It should be stable in both oxidised and reduced forms.

(5) The reduced state should be insensitive to oxygen.

(6) It should be readily immobilised.

(7) It should be non toxic for many applications e.g. _in vivo_ monitoring.

In 1984, the Cranfield-Oxford group (Cass et al., 1984) showed that a group of compounds (the ferrocenes) which largely satisfy most of the above criteria could be incorporated into a biosensor design for glucose. Glucose sensors were made by immobilizing glucose oxidase onto graphite foil chemically modified with 1,1'dimethylferrocene.

OPTICAL SENSORS

In an optical sensor, interaction of the biocatalyst with the target analyte brings about a change in the optical properties of the system. Depending on the particular device and the nature of the biological reaction the optical parameters measured can be absorbance, reflectance or luminesence. As such, optical sensors offer a number of potential advantages over devices based on electrochemical detection. These include:

(1) Lack of susceptibility to electrical interference.

(2) No reference electrode is required.

(3) Because the reagent phase does not need to be in physical contact with the transducer it is a simple matter to change the reagent phase.

(4) Optical probes can be made which operate on the principle of multi-wavelength monitoring. Thus it would be possible to monitor more than one species simultaneously.

Possible disadvantages, however, include interference from ambient light and long term stability of the reagent phase particularly of the light emitting components. An example of an optical probe is the pH sensor based on an immobilized dye, the optical properties of which are sensitive to the pH of the environment. This system has been extended into the development of optical pH probes for physiological applications by Peterson et al. (1980). Lowe et al. (1983) have reported an optoelectronic sensor for glucose based on glucose oxidase and immobilised bromocresol green.

An exciting area where optical sensors may prove particularly fruitful is in monitoring the reaction between an antibody and its respective antigen. Here, the antibody is immobilised onto the surface of an optically transparent waveguide, a glass or plastic plate or optical fibre, and interaction with the antigen is monitored by the absorption of light derived from multiple internal reflections within the wave guide (Sutherland, 1984).

THERMOMETRIC SENSORS

In theory, practically all chemical reactions can be monitored by measuring the enthalpy change associated with the particular chemical transformation. This rationale forms the basis of biosensors based on thermistor probes. This approach can be applied equally well to enzyme, multienzyme, cell and tissue based systems.

OTHER SENSORS

An alternative measuring principle is the exploitation of non specific ionic conductance. As we have seen, enzymes such as urease produce ionic species (NH_4^+, HCO_3^-) from neutral substrates. Thus such a reaction could be monitored by following the increase in solution conductance.

The use of piezo electric crystals for trace analysis of compounds such as anaesthetic gases, explosives and nerve gases has been reported in the literature (Ho et al., 1980). The idea of using a piezo electric crystal stems from the fact that the frequency of the crystal decreases as a function of weight increase, e.g. when volatile compounds adsorb to the surface. In order to obtain selectivity the crystal must first be coated with a suitable "substrate". Thus for example, carbowax is used to adsorb haloethanes. The same principle might also be utilized with enzymes acting as the adsorbing active layer.

Another area in which there is a rapidly growing interest is in the exploitation of semiconductor technology. The semiconductors which have been used in most of these studies are Field Effect Transistors (FET). Components such as these which are sensitive to ionic species or chemical agents are called ISFETS or CHEMFETS respectively. The most attractive features of semiconductor devices lies in their potential for miniaturisation and direct integration with microelectronics. The sensing portion of the semiconductor can be very small ($^\sim$1mm^2) and it should be possible to mount this point at the tip of a tiny catheter e.g. for in situ measurements in a patient's blood vessel. Caras and Janata (1980) described a probe consisting of a FET with immobilized β-lactamase activity for the enzymic detection of penicillin. Another group has studied the biological applications of gas sensitive semiconductor components having sensitivity for hydrogen, ammonia and hydrogen sulphide gas (Lundstrom, 1978). Semiconductor manufacturing technology is expected to play a major role in the development of a variety of biosensor types for the commercial market.

FERROCENE BASED AMPEROMETRIC SENSORS

The major impetus in our laboratories has been the exploitation of mediated electron transfer from reduced enzymes to electrode surfaces. Although reports have appeared in the literature describing direct electron transfer from an enzyme to a naked electrode, rates of transfer have tended to be slow and in the case of the dropping mercury electrode resulted in the denaturation of the enzyme. More promising results have been obtained where the natural redox partner for the enzyme has been replaced by the electrode. Organic metal electrodes comprising of N-methylphenazinium and the anionic radical tetracyano-4-quinodimethane have been used to effect electron transfer from the electrochemically active centre of flavin containing enzymes (Kulys et al., 1984). Similarly, chemically modified electrodes have proved useful in this respect (Cass et al., 1984).

The system commonly employed in our laboratory is based on 1,1-dimethylferrocene modified carbon electrodes. Ferrocene is a transition metal π arene complex which consists of an iron atom sandwiched between two cyclopentadienyl rings. In the biosensor configuration, electrochemically generated ferricinium ions, $Fe(Cp)_2^+$, act as oxidants for the immobilised enzymes. Once the ferricinium ion has been reduced, it can be reoxidised at the electrode surface by polarising the electrode and allowing current to flow. Thus, for example, the sequence of reactions occurring at the ferrocene modified glucose oxidase electrode can be represented as:

The low formal potential required to generate the ferricinium species (220mV vs Ag/AgCl$_2$) tends to minimize interferences and the fact that it does not react with oxygen makes this approach particularly suitable to clinical and fermentation monitoring where oxygen tension may vary.

Since the original 1984 publication the ferrocene based system has been extended to include a number of different enzymic activities. These are outlined in Table 1.

Electrodes routinely used in our laboratories are constructed as follows. Graphite foil supplied by either Union Carbide or Le Carbonne is used as the base sensor. Electrodes are constructed by cutting the foil into discs and sealing them into a glass tube with epoxy resin. Connection to the external circuit is made by a wire attached to the back of the electrode with silver Araldite. 1,1'-Dimethyl ferrocene (0.1M) in toluene is then applied to the surface of the electrode and dried in a stream of air. Finally, the enzyme is either covalently linked to the surface of the graphite using carbodiimide chemistry or held behind a suitable membrane.

This experimental approach brings out a number of desirable features in our sensors. First, the electron transfer reaction between the reduced enzyme and the ferricinium tends to be fast resulting in rapid response times in the sensors. Because of the low solubility of the ferrocene derivative in aqueous media the mediator is essentially confined to the electrode surface. Thus, such probes lend themselves to direct application. Finally, because the enzyme is physically trapped the sensors in question can be used more than once.

One of the most critical features affecting the usefulness of any biosensor is its long term stability which may be defined as the resistance to change in the slope of the calibration curve as a function of time. From Table 1 it is evident that the stabilities of our different sensors vary quite considerably. The factors which can influence the stability of a

Table 1. Performance of Some Biosensors Based on Ferrocene-Modified Electrodes

Substrate	Enzyme	Immobilisation	Range (mM)	Linearity (mM)	Response time (sec to 95%)	Half life in Continuous use at 30°C (h)
Glucose	Glucose oxidase	Covalent	<0.1 to 70	0 to 30	60 to 90	24 to > 600
Glucose	Quinoprotein glucose dehydrogenase	Covalent	<0.25 to 4.0	0 to 4	<30	36
Methanol	Methanol dehydrogenase	Entrapped	1×10^{-3} to 0.2	0 to 0.1	<20	1.5
Carbon Monoxide	CO oxidoreductase	Entrapped	$<2 \times 10^{-5}$ to $>7 \times 10^{-2}$	0 to 0.068	<10	≃ 6
L-Lactate	Lactate dehydrogenase	"	1 to 300	0 to 300	75	≃ 8
Amino Acid	Amino acid oxidase	"	<1 to 15	0 to 4	120	18
Glycollate	Glycollate oxidase	"	<1 to 20	0 to 7	180	3
Galactose	Galactose oxidase	"	<1 to 40	0 to 20	180	≃ 1.5
NADH	Lipoamide dehydrogenase	Covalent	2×10^{-3} to $>2 \times 10^{-2}$	-	<60	≃ 1.5

sensor are listed in Table 2. It must be stressed, however, that it is extremely difficult to predict the stability of a biosensor at the conceptual stage and it can only be successfully determined experimentally. A substantial amount of development work may be needed in some cases to achieve adequate stability for commercial exploitation. A final point which is particularly relevant to enzyme based sensors is that the apparent stability will depend very much on whether or not the sensor is tested at low or high substrate concentration with respect to the apparent Michaelis constant of the enzyme. Another point which emerges from Table 1 is that in most cases the kinetic properties of most enzymes change when included in the sensor configuration. This is particularly true of sensors based on glucose oxidase and L-lactate dehydrogenase. In both these examples, the $Km(app)$ of the immobilized enzyme has altered with respect to the soluble system, resulting in an extended linear response with respect to substrate concentration. The reasons for this observation are not yet clear but probably arise from an interplay of diffusional considerations, immobilisation chemistry, and interaction of the ferricinium ion and the reduced enzyme.

Apart from obvious clinical applications another area where biosensors will play an important role is in fermentation technology. Control of fermentation processes requires accurate and reliable information on variables such as substrate concentration, cell count, pH, pCO_2 etc. Without adequate knowledge of these parameters, it is impossible to optimise the growth conditions to give maximum yield of desired fermentation products.

We are at present investigating possible applications of our sensors to this area. For example, it is hoped that the ferrocene based glucose sensor can be used as a means of effectively controlling the rate of substrate feed to a batch fermentor. Initial experiments using the electrode as an on line sensor have indicated that it is suitable both in terms of response time and linearity. Different configurations such as the wall jet are now being investigated and evaluated.

Another important variable to monitor is the number of viable cells present in the fermentation broth. The development of such a sensor has been an extension of our earlier work with mediated fuel cells (Turner et al., 1982) where we noted that the current generated by bacteria in a phenazine ethosulphate (PES) mediated fuel cell was proportional to the concentration of organisms present in the cell. We have adapted the idea to produce a mediated amperometric sensor in which detection of the electro-active species (the mediator) takes place at a naked platinum electrode. Because of the instability of PES, other mediators have been tried and to date detection limits have been as low as 5×10^6 cells/ml.

Table 2. Factors Affecting the Stability of
an Enzyme Electrode

(1) The stability of the base sensor.

(2) The thickness of the enzyme layer.

(3) The mechanical stability of the trapped layer.

(4) Immobilization method.

(5) The total enzyme activity in the trapped layer.

(6) The stability of the trapped enzyme.

(7) Chemical conditions of use.

(8) Storage conditions.

The ultimate aim of this research is to interface the biosensor work with computer technology and create fermentation processes which are entirely under computer control. The work currently in progress in our laboratories is modelled around the BBC micro computer which in conjunction with the various probes mentioned above is being "trained" to monitor and respond to the changing environmental parameters associated with the particular fermentation.

Another area in which we are currently active is in developing enzyme linked immunoassays, based on the ELISA principle, but employing an electro-chemical detection mechanism as the final step. Exploitation of this work requires the construction of electrochemically suitable enzyme-antibody complexes by linking together electroactive "enzymes" with the antibody of interest. In the final configuration, the conjugates will be bound to affinity columns containing immobilised target antigen. The addition of sample containing soluble antigen will result in the displacement of a certain proportion of bound conjugate from the column. (The amount of liberated complex will be proportional to the concentration of free antigen in the original sample). The displaced antibody-antigen complex can then be monitored by tapping into the activity of the enzyme present in the conjugate. It will be possible by employing an electrochemical detection system to obtain rapid response times as well as good sensitivity. In order to be able to detect very small amounts of antigen by this technique it is important to use enzymes with extremely high turnover numbers in the construction of the conjugates. Enzymes which are particularly suited to this task are, for example, catalase (one of the fastest known enzymes) and the quinoprotein glucose dehydrogenase which has a turnover number of 320,000 molecules of glucose/min. The essential features of this technique are outlined in Figure 2.

The development of new biosensor devices will of course be facilitated by novel enzymes which lend themselves to incorporation into biosensor devices. A particularly useful class of enzyme are the quinoproteins. These enzymes, which include glucose dehydrogenase and non-NAD linked lactate dehydrogenase, use pyrollo-quinoline quinone (PQQ) as a prosphetic group. Unlike the flavin linked oxidases, PQQ shows complete insensitivity to oxygen, i.e. dioxygen will not act as a terminal electron acceptor for the reduced enzymes. PQQ will, however, react with artificial electron acceptors such as ferrocene. This class of enzyme is therefore suitable for biosensor design particularly where gross fluctuations in the local oxygen tension would otherwise create problems.

Sensors based on PQQ linked GDH and LDH have been constructed using ferrocene modified graphite as the base sensor. The performance of a GDH based sensor and a conventional GOD glucose sensor are compared below.

Enzyme	Immobil-isation	Range (mM)	Response time(s)
GOD	Covalent	0.1-70	60 to 90
GDH	Covalent	0.25-4.0	ˋ30

Although the sensors based on GDH do not show a large response with respect to substrate range it is hoped to remedy this by either finding a more suitable immobilisation chemistry or by increasing the enzyme loading. In this respect, we have shown that the range can be increased by 20mM by using immobilization techniques based on glutaraldehyde.

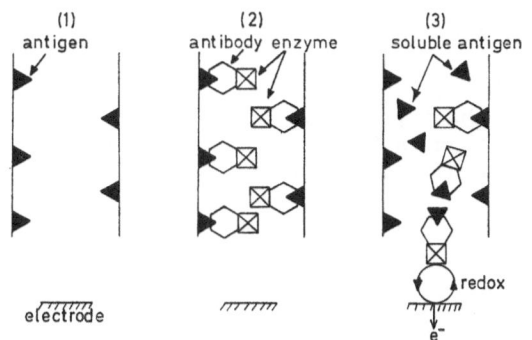

Fig. 2. Principles of an enzyme linked electrochemical immuno-assay. (1) Affinity column containing immobilised antigen. (2) Column charged with enzyme/antibody conjugate. (3) Addition of free antigen causes displacement of bound conjugate. This is then monitored by following the electrochemical activity of the enzyme.

CONCLUDING REMARKS

The work presented in this paper has in the main dealt with exploiting the catalytic activities of oxidases together with suitable mediator electrochemistry. As yet, we have not expanded our systems into multienzyme configurations i.e. where the product of one enzyme acts as the substrate of the other although we have demonstrated the use of the ferrocene/glucose oxidase sensor to measure serum creatine kinase (Davis, in press). By employing such an approach it is hoped to be able to increase the substrate range of our present sensors and possibly even increase their limit of detection.

Promising related research activity concerns the development of electro-chemical techniques for monitoring whole cell metabolism. In this respect we are particularly interested in using the cyanobacteria because they exhibit three distinct metabolic activities, namely photosynthesis, oxidative phosphorylation and nitrogen fixation. By using specific mediators and suitable electrode configurations it is hoped to be able to monitor each of these metabolic processes individually. Because it is known that these organisms are particularly sensitive to the presence of herbicides it should be possible to extend this work into the design of whole cell based sensors which would be particularly suited to pollution control.

REFERENCES

Calvot, C., Berjonneau, A.M. Gelif, G. and Thomas, D., 1975, FEBS letters, 59, p 258.
Caras, S. and Janata, J., 1980, Anal. Chem., 52, p 1935.
Cass, A.E.G., Davis, G., Francis, G.D., Hill, H.A.O., Aston, W.J., Higgins, I.J., Plotkin, E.V., Scott, L.D.L. and Turner, A.P.F., 1984, Anal. Chem., 56, p 667.
Clark, L.C. and Duggan, G.A., 1982, Diabetes Care, 5, p 174.
Guilbault, G.G. and Shu, F.R., 1972, Anal. Chem., 44, p 2161.
Ho, M.H., Guilbault, G.G. and Reitz, B., 1980, Anal. Chem., 52, p 1489.
Karube, I., Misuda, S. and Suzuki, S., 1979, Eur. J. Appl. Microbiol. Biotechnol. 7, p 343.

Kulys, J.J., Pocius, A.K. and Cenas, N.K., 1984, Biochem. Bioenerg. 12, p 583.

Lowe, C.R., Goldfinch, M.J. and Lias, R.J., 1983, Biotech 83, Online Publications, Northwood, p 633.

Lundstrom, I., 1978, Physica Scripta, 18, p 424.

Nagy, G., Von Storp, L.H. and Guilbault, G.G., 1973, Anal. Chim. Acta. 63, p 443.

Nanjo, M. and Guilbault, G.G., 1974, Anal. Chim. Acta., 73, p 367.

Nilsson, H., Akerlund, A.C. and Mosbach, K., 1973, Biochim. Biophys. Acta., 320, p 529.

Papariello, G.J., Mukherji, A.K. and Slearer, C.M., 1973, Anal. Chem., 45, p 790.

Peterson, J.I., Goldstein, S.R., Fitzgerald, R.V. and Buckhold, P.K., 1980, Anal. Chem. 52, p 864.

Siddiqi, I.W., Clin. Chem., 1982, 25, p 1962.

Sutherland, R.M., Dahne, C., Place, T.F. and Ringrose, A.S., 1984, Clin. Chem., 30, p 1533.

Turner, A.P.F. and Pickup, J.C., 1985, Biosensors 1, p 85.

Turner, A.P.F., Ramsay, G. and Higgins, I.J., 1982, Biochem. Soc. Trans. 11, p 445.

Williams, D.L., Doig, A.R. and Korosi, A., 1970, Anal. Chem., 42, p 118.

THE D-ALANYL-D-ALANINE-CLEAVING PEPTIDASES AND β-LACTAMASES.   STRUCTURE

AND MECHANISM

Jean-Marie Ghuysen

Service de Microbiologie, Faculté de Médecine
Université de Liège, Institut de Chimie, B6
B-4000 Sart Tilman (Liège), Belgium

INTRODUCTION

   Enzyme kinetics, amino acid and nucleotide sequencing, oligonucleotide-
directed mutagenesis, protein crystallography and X-ray diffraction,
computer molecular graphics and theoretical chemistry contribute much to
the unravelling of the mechanisms of enzyme catalysis.   The present paper
summarizes our state of knowledge concerning several peculiar bacterial
enzymes, namely the active-site serine and active-site Zn^{++} DD-peptidases
and β-lactamases.

STRUCTURE AND BIOSYNTHESIS OF THE BACTERIAL WALL PEPTIDOGLYCAN

   Bacteria are surrounded by a plasma membrane that contains many enzymes
and fulfils many essential functions.   Because of the large concentration
gradient between the inside and the outside of the plasma membrane, water
has a strong tendency to flow inward.   To preserve the bacteria against
osmotic disruption, a wall of high tensile strength outside the plasma
membrane has evolved.   The wall keeps the bacterial cell alive under
ordinary hypotonic environmental conditions, thanks to the presence of a
specialized heteropolymer called peptidoglycan.

   Peptidoglycans are all built on the same general pattern (Ghuysen, 1968;
Ghuysen, 1977).   This cell supporting network structure consists of linear
strands of alternate, β,1-4 linked N-acetylglucosamine (GlcNAc) and
N-acetylmuramic acid (MurNAc) pyranoside residues.   Every D-lactyl group
of N-acetylmuramic acid is substituted by a tetrapeptide L-Ala-γ-D-Glu-L-
Xaa-D-Ala where, most often, L-Xaa is a diamino acid such as L-Lys or
meso-A_2pm.   Peptide units substituting adjacent glycan chains are
covalently linked together by means of "bridges".   These bridges extend
between the C-terminal D-Ala of a peptide and the ω amino group of the
L-Xaa residue of another peptide.   The bridges either consist of a direct
N^{ω}-(D-alanyl)-L-Xaa peptide bond [for example, D-Ala-(L)-meso-A_2pm in the
Gram-negative bacteria (Figure IA)] or are mediated via a single additional
amino acid residue [for example, N^{ε}-D-isoasparaginyl-L-lysine in
Streptococcus faecium and Lactobacillus acidophilus (Figure 1B)] or an
intervening peptide [for example, N^{ε}-(D-alanyl-pentagylcyl)-L-Lys in
Staphylococcus aureus (Figure 1C)].

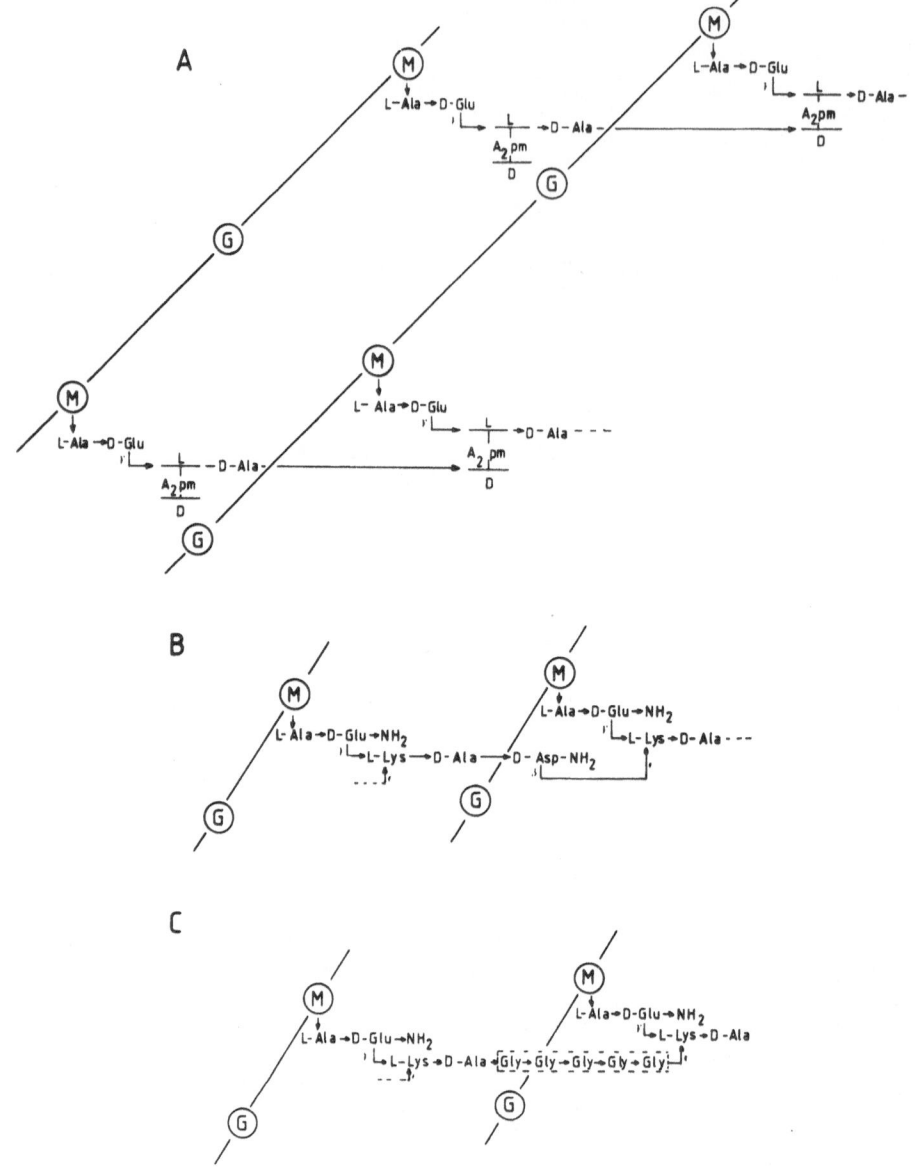

Fig. 1. Wall peptidoglycans in Escherichia coli (A), Streptococcus faecium and Lactobacillus acidophilus (B), and Staphylococcus aureus (C). G = N-acetylglucosamine; M = N-acetylmuramic acid.

Bacterial wall peptidoglycan synthesis is a multi-stage process which involves 1) the manufacture, in the cytoplasm, of the two nucleotide precursors UDP-GlcNAc and UDP-MurNAc-L-Ala-γ-D-Glu-L-Xaa-D-Ala-D-Ala; 2) the assembly, from these precursors, of β,1-4 linked GlcNAc-MurNAc-L-Ala-γ-D-Glu-L-Xaa-D-Ala-D-Ala on a C_{55} polyisoprenoid alcohol phosphate carrier and the transfer of the disaccharide peptide units through the plasma membrane; and 3) the incorporation of the newly synthesized disaccharide peptide units into the preexisting wall peptidoglycan. In those bacteria where the peptidoglycan interpeptide bridges consist of one or several additional amino acid residues (Figure 1B-1C), extension of the ω amino terminus of the L-Xaa residue takes place sometimes at the level of the UDP-MurNAc-peptide precursor, most often at the level of the lipid intermediate. Peptidoglycan synthesis thus requires formation of many glycosidic and peptide (amide) bonds.

ENZYME-CATALYSED PEPTIDE BOND FORMATION DURING BACTERIAL WALL PEPTIDO-GLYCAN SYNTHESIS

Enzyme-catalysed peptide bond formation obeys the following rules: it is made at the expense of ATP; it proceeds via conversion of the carbonyl donor cosubstrate to an ester-linked intermediate; it is achieved by attack of the carbonyl carbon of the ester intermediate by the amino acceptor co-substrate.

In the well-known process of DNA template-directed protein synthesis, each amino acid destined to form the polypeptide chain is first converted by a specific synthetase at the expense of ATP, to an aminoacyl-tRNA where the aminoacyl moiety is ester-linked to the 3'-OH ribosyl moiety of the 3' terminal adenosyl group of the tRNA (Figure 2). A ribosome-bound peptidase (peptidyl transferase) then catalyses attack on the carbonyl carbon of the ester bond of the acly tRNA fixed on the donor site of the ribosome, by the N̈H$_2$ group of the aminoacyl-tRNA fixed on the acceptor site of the same ribosome (Figure 3). As a result of the transfer reaction, the donor site is vacated by the deacylated tRNA and reoccupied by the peptidyl-tRNA which is translocated from the ribosome acceptor site with the codon (on the mRNA)-anticodon (on the tRNA) interaction remaining intact. The relative movement of the ribosome towards the 3' end of the mRNA exposes the next codon at the vacant acceptor site, permits binding of the next aminoacyl-tRNA at this site and further action of the peptidyl transferase. Through a repetition of this process, the nascent polypeptide chain grows by extension at its carboxy-terminus until a terminator codon (UAA, UAG or

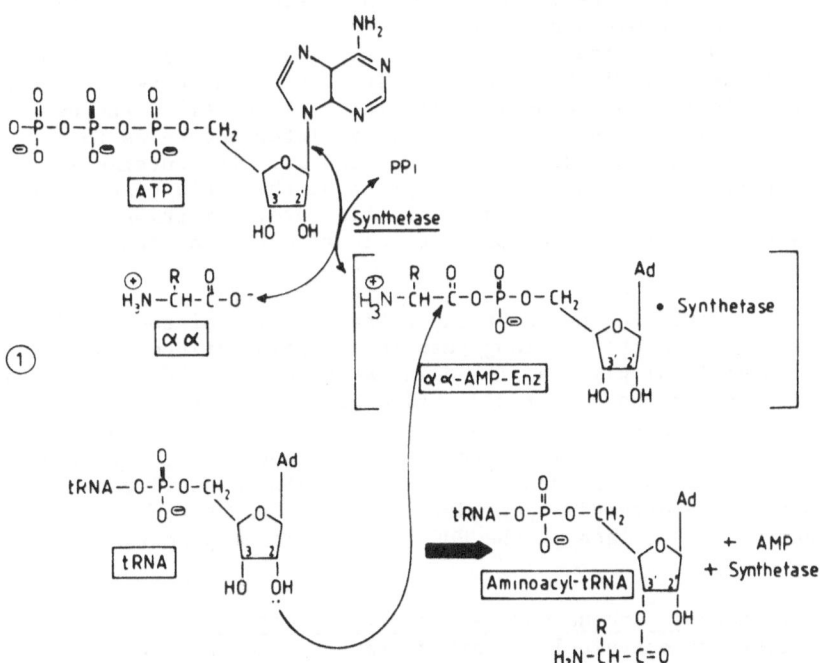

Fig.2. Synthetase-catalysed formation of amino-acyl tRNA.

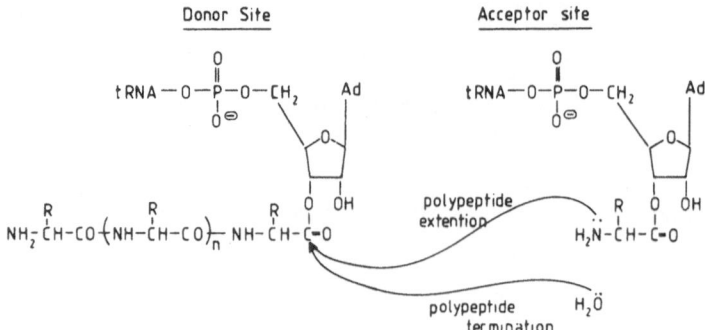

Fig. 3. Peptide bond formation and polypeptide
termination in DNA-template-directed
protein synthesis on ribosome.

UGA) in the mRNA appears at the ribosome acceptor site which therefore
remains empty. The peptidyl transferase then uses H_2O as nucleophilic
acceptor of the transfer reaction (Figure 3) causing detachment of the
complete polypeptide chain at the carboxy-terminus. Hence, depending on
whether an amino group is available or not, the same peptidyl transferase
catalyses peptide bond formation or hydrolysis of the acyl-tRNA.

The principles defined above apply to the enzyme-catalysed peptide
(amide) bond formation during wall peptidoglycan synthesis. However, four
distinct mechanisms are involved, none of them is ribsome-mediated and the
ester-linked intermediate is not always an aminoacyl-tRNA (for a review,
see Ghuysen, 1977).

1) The manufacture of the UDP-MurNAc pentapeptide precursor relies on the
action of several specific, water-soluble ligases (ADP). They catalyse
synthesis of a D-Ala-D-Ala dipeptide, addition of L-Ala to the precursor
UDP-MurNAc, addition of D-Glu to UDP-MurNAc-L-Ala, addition of L-Xaa to
UDP-MurNAc-L-Ala-γ-D-Glu and addition of the preformed D-Ala-D-Ala to UDP-
MurNAc-L-Ala-γ-D-Glu-L-Xaa. The D-Ala-D-Ala ligase (ADP) has been investigated in some detail. It has been proposed that i) an ordered sequence of
D-Ala binding in the enzyme active site takes place (first the donor; then
the acceptor); ii) an ester-linked D-alanyl intermediate is formed at the
enzyme donor site at the expense of ATP; and iii) the intermediate undergoes nucleophilic attack by the amino group on the D-Ala cosubstrate bound
at the enzyme acceptor site. Hence, in this case, formation of the ester
intermediate and peptide bond formation occurs within a single enzyme active
site (Figure 4).

2) Extension of the side chain at the L-Xaa position of the peptides (in
those bacteria where the peptidoglycan interpeptide bridges consist of one
or several additional glycine or L-amino acid residues (Figure 1C)) involves
conversion of the amino acid residue(s) destined to form the interpeptide
bridges to aminoacyl-tRNAs. Ribosomes, however, are not involved and the
accuracy of the process entirely relies on the specificity profiles of the
peptidyl transferases for both the aminoacyl-tRNA carbonyl donor and the
amino group on the extending peptide. Depending on the bacterial species,
these reactions take place on the UDP-MurNAc peptide precursor or the
lipid intermediate.

3) In Streptococcus faecium and Lactobacillus faecalis, the interpeptide
bridge consists of one single isoasparaginyl residue having the D-
configuration (Figure 1B). D-aspartate is first converted to β-D-aspartyl

70

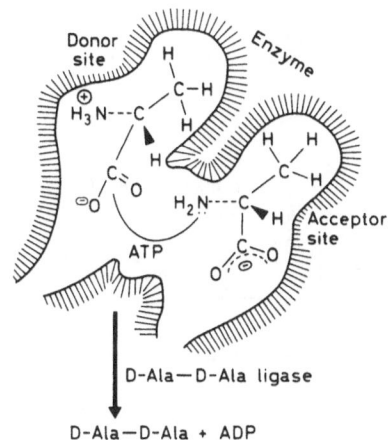

Fig. 4. Putative mechanism of the
synthesis of D-Ala-D-Ala
by the D-Ala:D-Ala ligase
(ADP).

phosphate at the expense of ATP and the β-D-aspartyl moiety is then
transferred to the ε amino group of the L-lysine residue. The mechanism
occurs at some stage of the lipid cycle.

4) Finally, closure of the peptidoglycan interpeptide bridges is made by
transpeptidation. Transpeptidation, as illustrated in Figure 5A for a
peptidoglycan where the interpeptide bridge is a direct D-Ala-(L)-meso-A_2pm
linkage, is a reaction through which the C-terminal D-Ala-D-Ala peptide bond of
a peptide unit, acting as carbonyl donor, is attacked by the ω amino group

peptide unit, acting as carbonyl donor, is attacked by the ω amino group
at the L-Xaa position of another peptide, acting as nucleophilic acceptor.
The C-terminal D-Ala is released from the carbonyl donor and a new N^ω-D-
alanyl-L-Xaa peptide bond is formed through which the two peptide cosub-
strates are linked to each other. Transpeptidation does not require any
input of energy but it is made at the expense of a preformed D-Ala-D-Ala
peptide bond the synthesis of which has been catalysed by a ligase (ADP)
at an early stage of the biosynthetic pathway with accompanying hydrolysis
of one molecule of ATP. As shown below, catalysis involves formation from
the D-Ala-D-Ala terminated peptide of an ester-linked acyl enzyme intermediate (with release of D-Ala) at the enzyme donor site and subsequent
transfer of the acyl moiety to the ω amino group of the other peptide bound
at the enzyme acceptor site.

DD-PEPTIDASES AND WALL PEPTIDOGYLCAN EXPANSION

Transpeptidation reactions lead to reticulation and insolubilization
of the peptidoglycan. They are essential steps in wall biosynthesis.
However, completed wall peptidoglycans always contain varying amounts of
tetrapeptide ∿L-Ala-γ-D-Glu-L-Xaa-D-Ala units (and sometimes tripeptide
∿L-Ala-γ-D-Glu-L-Xaa units). They occur either in the form of uncrosslinked monomers or at the carboxy-terminus of peptide dimers or oligomers.
The tetrapeptides are either the products of a carboxypeptidation reaction
on pentapeptide units (Figure 5B) or they originate by cleavage of N^ω-D-
alanyl-L-Xaa interpeptide bonds previously made by transpeptidation
(Figure 5C). In Escherichia coli and other Gram-negative bacteria, crosslinked peptides may be cleaved at a peptide bond between a D-Ala residue

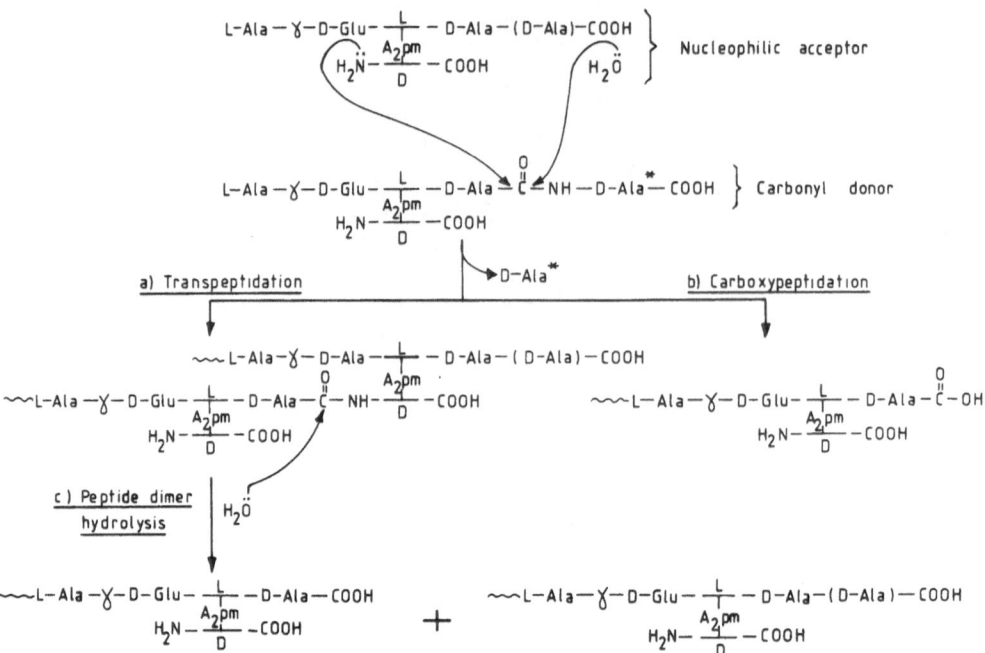

Fig. 5. Transpeptidation (A), carboxypeptidation (B) and peptide dimer
hydrolysis (C) during wall peptidation metabolism in
<u>Escherichia coli</u>.

and another D-residue bearing a free carboxylate (Figure 5C). In this case,
opening of the peptide bridges (which may cause peptidogylcan solubilization) is also the result of a carboxypeptidation reaction.

    All the enzymes which catalyze transpeptidation and carboxypeptidation
reactions in wall peptidoglycan metabolism have in common a unique optical
specificity: the scissile peptide bond extends between two D centres in
α position to a free carboxylate. For this reason, they are called DD-
peptidases.

    Identification of DD-peptidases has been made possible thanks to their
propensity to be inactivated by β-lactam compounds (penicillins, cephalo-
sporins, monobactams) with which they form adducts sufficiently stable to
be separated by SDS gel electrophoresis. If a radioactive β-lactam is
used, fluorography of the gels permits visualization of the labelled
proteins (Spratt, 1983). The efficacy of β-lactam binding, however, depends
largely on the DD-peptidase and the β-lactam considered. The process is
both enzyme and β-lactam specific.

    In <u>Escherichia coli</u>, wall peptidoglycan expansion is made by direct
incorporation of newly synthesised disaccharide peptide units to the "old"
peptidoglycan by transglycosylation at the level of the glycan chains and
by transpeptidation at the level of the peptide moiety. For this purpose,
<u>E. coli</u> possesses in the cytoplasmic membrane, at least seven penicillin-
binding DD-peptidases with relative molecular masses ranging between 91,000
to 40,000.

    Genetic analysis (Spratt, 1983; Broome-Smith et al., 1985) of the
role of each of these proteins in the lethal effect of β-lactams has shown
that the lower-Mr proteins 4, 5 and 6 are not essential for bacterial

growth and are not of prime importance for the killing action of the β-lactams. Proteins 4, 5 and 6 catalyse both carboxypeptidation and trans-peptidation reactions with varying efficacy. In all likelihood, each of them possesses one single active site that performs these two types of DD-peptidase activity.

The E. coli proteins 1A, 1B, 2 and 3 are essential enzymes and β-lactams exert their lethal effect by inactivating one or more of these killing targets (Spratt, 1983; Broome-Smith, 1985). Inactivation of protein 3 causes inhibition of cell division and formation of filamentous cells. Inactivation of protein 2 results in the growth of E. coli as spherical cells. Inactivation of either protein 1A or protein 1B is not lethal under laboratory conditions but double mutants, lacking both protein 1A and protein 1B activity, are non viable. Proteins 1A and 1B probably have compensatory roles in peptidoglycan synthesis. Inactivation of both proteins 1A and 1B causes rapid cell lysis of E. coli. Proteins 1A, 1B, 3 and possibly 2 catalyse the in vitro synthesis of crosslinked peptido-glycan from lipid-linked peptidoglycan precursors (unfortunately with a very low turnover number of about 0.01 s^{-1} or less (Nakagawa et al., 1979; Suzuki et al., 1980; Ishino et al., 1980; Ishino and Matsuhashi, 1981). These enzymes are therefore bi-functional. Each of them possesses two active sites that perform transglycosylase and transpeptidase activity, respectively. Inactivation of the transpeptidase site by β-lactams does not prevent the transglycosylase site from functioning but, conversely, transpeptidase activity appears to require either prior or concomitant transglycosylation.

As derived from nucleotide sequences, the primary structures of the E. coli proteins 1A, 1B, 3 and 5 have been established (Broome-Smith et al, 1985; Broome-Smith et al, 1983; Nakamura et al., 1983) (Figure 6). Proteins 1A and 1B show substantial similarities. They are closely homo-logous even if major rearrangements have occured during their divergence. In contrast, proteins 1A/1B, protein 3 and protein 5 show no sequence similarity to one another except for two small regions. One region, shown in Figure 6, includes the sequence (Gly or Ala)-Ser-Xaa-Xaa-Lys with the serine residue occurring at position 44 in protein 5, 307 in protein 3, 510 in protein 1B and 465 in protein 1A. In each case, this serine residue has been shown to be the binding site of radioactive penicillin (Keck et al., 1985). The second region of sequence similarity (not shown in Figure 6) occurs at a variable distance of 17-38 residues to the amino-terminal

E.coli Penicillin binding proteins		
	1 A	AsnArgAlaThrGlnAlaLeuArgGlnValGly Ser Asn Ile Lys Pro Phe Leu Tyr Ala Ala Ala Met
	1 B	AsnArgAla MetGlnAlaArgArgSer Ile Gly Ser Leu Ala Lys Pro Ala Thr Tyr Leu Thr Ala Leu
	3	AsnArgThr Ile ThrAspVal PheGluProGly Ser Thr Val Lys Pro Met Val Val Met Thr Ala Leu
	5	GluGlnAsnAla AspVal ArgArgAspProAla Ser Leu Thr Lys Met Met Thr Ser Tyr Val Ile Gly
β-lactamases RTEM		Ser Phe Arg Pro Glu Glu Arg Phe Pro Met Met Ser Thr Phe Lys Val Leu Leu Cys Gly Ala Val Leu
ampC		Pro Val Thr Gln Gln Thr Leu Phe Glu Leu Gly Ser Val Ser Lys Thr Phe Thr Gly Val Leu Gly Gly
R61 DD-peptidase		Val Gly Ser Val Thr Lys Ser Phe Ser Ala Val Val Leu Leu

Fig. 6. Amino acid sequences around the active-site serine residue in DD-peptdases and β-lactamases.

site of the active serine residue (Broome-Smith et al., 1985), Enzymes that recognize the same inhibitor often have little primary structure similarity except in those segments of the polypeptide chains that form the inhibitor binding sites. Most likely, the two homologous regions observed in proteins 1A, 1B, 3 and 5 are part of the penicillin-sensitive, DD-peptidase active site.

The active-site serine occurs close to the amino-terminus in protein 5 and towards the middle of proteins 1A, 1B and 3 (Figure 6). This observation is consistent with the localization of the β-lactam insensitive peptidoglycan transglycosylase site at the amino-terminus domain and the β-lactam-sensitive transpeptidase site at the carboxy-terminus domain of proteins 1A, 1B and 3. As a corollary of this, the amino terminal 240 amino acids have been removed from protein 3 without loss of the penicillin-binding activity of the protein (but with considerable loss of stability) (Hedge and Spratt, 1984).

The above picture is specific to E. coli. As revealed by [^{14}C] penicillin binding experiments, each bacterial species appears to possess its own assortment of membrane-bound, penicillin-sensitive DD-peptidases. The DD-peptidases present in a single bacterial cell catalyse distinct reactions, fulfil distinct cellular functions, do not exhibit the same degree of essentiality and show varying sensitivity to β-lactam compounds. Moreover, in Gram-positive bacilli, uncrosslinked glycan chains consisting of multiple disaccharide units, grow by transglycosylation on the lipid carrier. This nascent peptidoglycan then undergoes insolubilization by peptide crosslinking between new chains and between new and old chains. Transglycosylase activity and transpeptidase activity are probably carried out by distinct proteins.

The DD-peptidases form a very large group of enzymes which differ from each other in their physical, molecular and enzymatic properties. The DD-peptidases also belong to at least two mechanistically distinct group of enzymes. In addition to the penicillin-sensitive, active-site serine DD-peptidases mentioned above, at least one penicillin-insensitive DD-peptidase is known. This DD-peptidase does not behave as a penicillin binding protein, it does not catalyse transpeptidation reactions and it functions only as a strict carboxypeptidase (peptidoglycan hydrolase). It is a metallo-DD-peptidase, the catalytic activity of which depends on the presence of a Zn^{++} ion cofactor in the enzyme active site. The Zn^{++} DD-peptidase is described in some detail in the last section of this article.

THE β-LACTAMASES

Bacteria have built up very effective lines of defense against action of the β-lactam antibiotics. One of them has been the rapid spread of genes coding for various types of β-lactamases. β-lactamases are enzymes which effectively hydrolyze the endocyclic amide bond of the β-lactam ring, converting the β-lactam antibiotics into biologically inactive metabolites (Figure 7).

The genetic information for β-lactamase can be located on two types of replicons, the chromosome and the plasmids, and one essential character-istic of the β-lactamase genes is mobility (Richmond et al., 1980). A given type of β-lactamase can be found among distinct bacterial species. β-lactamase genes may be plasmid-linked in some variants of a bacterial species while being chromosomal in others. Occasionally, bacteria may carry two β-lactamase genes at once, one chromosomal and the other plasmid-linked. All these observations imply mechanisms of gene transfer both from one genetic location to another and between bacterial species.

74

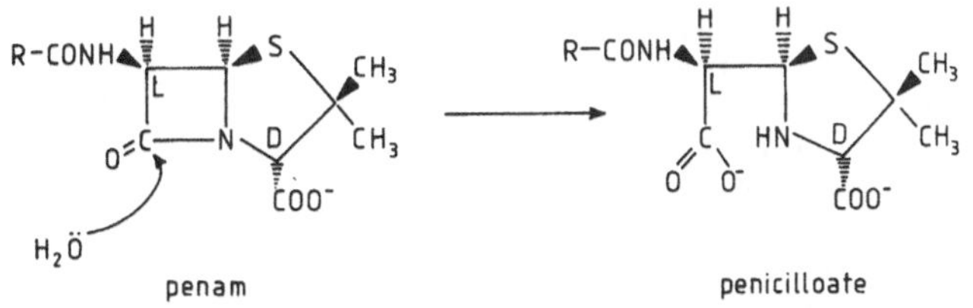

Fig. 7. β-Lactamase-catalysed hydrolysis of pencillin. Note the disposition of the carbon atoms with L and D configurations.

Many conjugative plasmids possess their own machinery to ensure their transfer directly from one cell to another. This mode of transfer which involves mating between a donor cell and a recipient cell applies, especially but not exclusively, to Gram-negative bacteria. In addition, plasmids are small enough to be incorporated in the head of a bacteriophage through which they can be transferred to another bacterial cell, a mode of transfer which is important among Gram-positive bacteria. Finally, β-lactamase genes are integrated into specialized DNA fragments, called transposons. Transposons are incapable of independent replication and have to rely on functional replicons (chromosome or plasmid) for replication. But they can move among various replicons and to various sites within a same replicon, independently of normal generalized recombination.

β-lactamase genes are mobile and bacteria can create new β-lactamases by recombination of existing genes. Such a system has high evolutionary potential but it can be deeply influenced by the environmental conditions. Today, as a consequence of the massive use of β-lactam antibiotics, almost every species of bacteria has members that have β-lactamase genes and more than 80 different types of β-lactamases have been detected. β-lactamases are widely diverse enzymes of polyphyletic origin (Ambler, 1980).

The recent advances achieved by gene analysis and the establishment of amino acid sequences and mechanistic properties has led to the recognition of distinct classes of β-lactamases. Labelling the β-lactamases with poor substrates or mechanism-based inactivators (see below) and the establishment of the amino acid sequences of the active binding sites have led to the conclusion that many β-lactamases are serine enzymes.

The RTEM β-lactamase of E. coli, the β-lactamase I of Bacillus cereus 569/H, and the β-lactamases of Staphylococcus aureus PC1 and Bacillus licheniformis 749/C are serine β-lactamases of class A (Ambler, 1980). These "penicillinase-type" β-lactamases have similar molecular masses of about 29,000 but varying isoelectric points (from pH 5.4 to 9.0), substrate profiles and kinetic parameters. In spite of these differences, they all match remarkably well throughout their amino acid sequences and only a few gaps need to be postulated to obtain optimal matching. A peptide fragment of the RTEM β-lactamase containing the active-site serine 45 is shown in Figure 6. Although the β-lactamases of class A have changed through evolution, the amount of similarity is still so great that they are clearly related in an evolutionary sense. They all probably derive from a common ancestor gene. Oligonucleotide-directed mutagenesis has been used to invert the Ser-Thr dyad (containing the active-site serine residue) to a Thr-Ser dyad. The resulting protein had no catalytic activity (Dalbadie-McFarland et al., 1982).

The chromosome-coded ampC β-lactamase of E. coli (a cephalosporinase-type enzyme) and the β-lactamases of Pseudomonas aeruginosa 18S and Enterobacter cloacae P99 are also serine enzymes. A peptide fragment of the ampC β-lactamase containing the active-site serine 61 is shown in Figure 6. That the ampC β-lactamase had to be assigned as a distinct class came from the established nucleotide sequence of the corresponding cloned gene (Jaurin and Grundström, 1981). The ampC β-lactamase completely lacks primary structure homology with the β-lactamases of class A and is the prototype of the β-lactamases of class C. Similarly, the established sequence of peptide fragments containing the active-site serine strongly suggest that the β-lactamases of P. aeruginosa (Knott-Hunziker et al., 1982) and E. cloacae (Joris et al., 1984) are also members of class C. The same conclusion probably appies to the β-lactamases of many enterobacteria (Bergström et al., 1982) as shown by hybridization experiments involving a ^{32}P-labelled fragment of DNA encoding for the major part of the ampC β-lactamase gene.

In the β-lactamases of class A and C, the active-site serine is located close to the amino terminus of the polypeptide chain (Figure 6) and has a phenylalanine and a lysine residue similarly situated relative to the active serine (Phe-Xaa-Xaa-Xaa-Ser-Xaa-Xaa-Lys) suggesting that, perhaps, these Phe and Lys residues play a role in catalysis. β-lactamases of class A and C have distinct origins but have evolved to a same mechanism of β-lactam ring hydrolysis. Interestingly, the sequence Ser-Xaa-Xaa-Lys has also been found so far at the active site of all the penicillin-binding serine DD-peptidases (including the exocellular DD-peptidase of Strepto-myces R61 (Frere et al., 1976)), but with the apparent exception of the DD-peptidase of Actinomadura R39 (Duez et al., 1981).

Contrary to the β-lactamases of class A and C, the β-lactamase II which is manufactured by B. cereus (together with the serine β-lactamase I of class A) (Ambler, 1980) and the β-lactamase of Pseudomonas maltophilia LD1275 (Bicknell et al., 1985) require a metal ion cofactor for activity. The two naturally occurring β-lactamases are zinc enzymes but Co^{++} in the β-lactamase II and Ni^{++} in the β-lactamase LD can replace Zn^{++}. The β-lactamase II has been assigned as a class B. Most likely, the β-lactamase LD belongs to another structural class of metallo β-lactamases.

CHYMOTRYPSIN AS MODEL OF THE SERINE DD-PEPTIDASES AND SERINE β-LACTAMASES

The usual serine peptidases such as chymotrypsin catalyse transfer of the electrophilic group $R^1-\overset{O}{\overset{\|}{C}}$ from amide ($R^1-\overset{O}{\overset{\|}{C}}-NH-R^2$) or ester ($R^1-\overset{O}{\overset{\|}{C}}-O-R^2$) carbonyl donors to a nucleophilic acceptor via formation of an active-site serine ester linked acyl ($R^1-\overset{O}{\overset{\|}{C}}-O$) enzyme (Fersht, 1984). The same mechanism applies to cysteine peptidases such as papain in which case a cysteine ester linked acyl enzyme is formed.

Giving E = enzyme; D = carbonyl donor; HY = nucleophilic acceptor; E·D = Michaelis complex; E-D* = acyl enzyme; P_1 = leaving group of the enzyme acylation step, i.e. NH_2-R^2 or $OH-R^2$; P_2 = second reaction product, i.e. R^1-COOH; K = dissociation constant of E·D; and k_{+2} and k_{+3} = first-order rate constants, the general reaction catalysed is

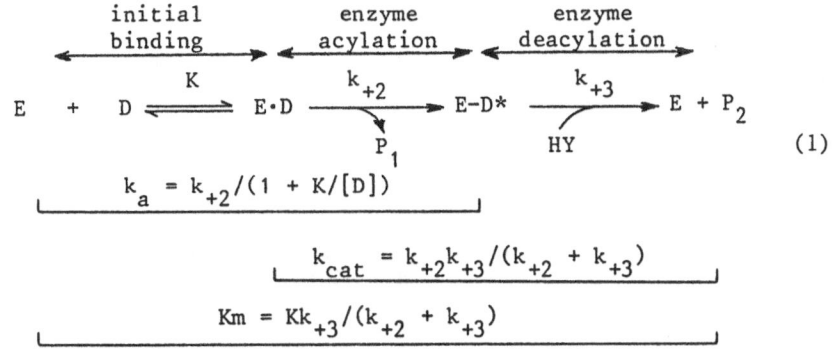

$$k_a = k_{+2}/(1 + K/[D])$$

$$k_{cat} = k_{+2}k_{+3}/(k_{+2} + k_{+3})$$

$$Km = Kk_{+3}/(k_{+2} + k_{+3})$$

It is assumed that the enzyme is present in catalytic amounts ($[D] \gg [E]$) and complex E·D is in rapid equilibrium with free enzyme and free carbonyl donor substrate. Constant k_a is the pseudo-first order rate constant of acyl enzyme formation at a given concentration of carbonyl donor substrate. Constants K_{cat} and Km are the usual kinetic parameters derived from Lineweaver-Burk plots $1/v$ _versus_ $1/[D]$. The ration k_{+2}/K (or k_{cat}/Km) is the second-order rate constant of enzyme acylation.

After a certain time (and for $[D] \gg E$), the system reaches a steady state where the concentrations of E, E·D and E-D* remain stable. At the steady state of the reaction, the ratio of total enzyme (E_0) to acyl enzyme ($[E-D*]_{ss}$) is

$$E_0/[E-D*]_{ss} = (k_a + k_{+3})/k_a = 1 + (k_{+3}/k_{+2}) + (k_{+3}K/k_{+2}[D]) \quad (2)$$

and the time which is necessary for the acyl enzyme to reach a certain percentage of its steady state level is

$$t = - \ln (1 - \frac{E-D*}{[E-D*]_{ss}})/(k_{+3} + k_a) \quad (3)$$

Thus, for example, for $K_{+2} = k_{+3} = 10 \text{ s}^{-1}$, $K = 1$ mM and $[D] = 10$ mM, the reaction reaches 99% of its steady state level after 0.24 s and at the steady state, the reaction proceeds with 47% of total enzyme being present as acyl enzyme.

From equation (2), it follows tht the higher the K_{+3}/k_{+2} ratio value, the lower the proportion of total enzyme which is present as acyl enzyme at the steady state of the reaction. For $k_{+3} \gg k_{+2}$, no acyl enzyme accumulates, formation of acyl enzyme is the rate determining step in catalysis and K_{cat} and Km simplify to $k_{cat} = k_{+2}$ and $Km = K$. On the contrary, for $k_{+3} \ll K_{+2}$, breakdown of acyl enzyme is rate determining, $k_{cat} = k_{+3}$, $KM = Kk_{+3}/k_{+2}$ and the ratio of total enzyme to acyl enzyme at the steady state is

$$E_0/[E-D*]_{ss} = 1 + (Kk_{+3}/k_{+2}[D]) \quad (4)$$

Assume now that the reaction between a serine peptidase and a carbonyl donor substrate involves two competing nucleophiles HY^1 and HY^2 acting at the level of the acyl enzyme. Reaction (1) becomes a branched pathway

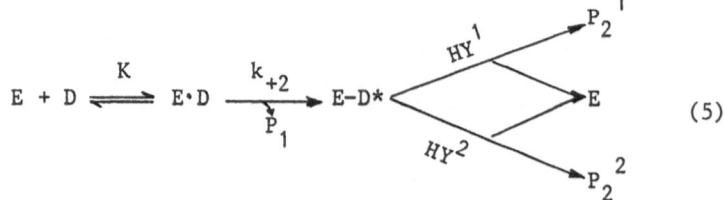

$$E + D \xrightleftharpoons{K} E{\cdot}D \xrightarrow[\substack{\downarrow \\ P_1}]{k_{+2}} E{-}D^* \begin{array}{c} \overset{HY^1}{\nearrow} P_2^1 \\ \to E \\ \underset{HY^2}{\searrow} P_2^2 \end{array} \qquad (5)$$

Partitioning of the enzyme activity depends on the competition between HY^1 and HY^2. If in the presence of HY^1 alone, the acyl enzyme accumulates at the steady state, increasing concentrations of HY^2 may increase the rate of acyl enzyme breakdown and hence may increase k_{cat} until the rate of enzyme acylation (which is supposed to be unmodified) becomes rate determining (i.e. k_{cat} becomes equal to k_{+2}) (Figure 8B). Alternatively, if in the presence of HY^1 alone, the acyl enzyme does not accumulate, partitioning still occurs but increasing concentations of HY^2 cannot increase k_{cat} (Figure 8A).

The usual serine peptidases are hydrolytic enzymes ($HY = H_2O$; $P_2 = R^1{-}COOH$). But L-alanine amide at a 1 M concentration, for example, can act as an alternate nucleophile. Both hydrolysis and aminolysis occur and the reaction products are $R^1{-}COOH$ and $R^1{-}CONH{-}CH(CH_3){-}CONH_2$. Similarly, the use of an alternate nucleophile stronger than water, for example a 1 M solution of neutral hydroxylamine, to speed up the rate of enzyme deacylation can be an effective means for releasing an active enzyme from a long lived acyl enzyme with formation of the hydroxamate $R^1{-}CO{-}NHOH$.

Serine peptidase-catalysed peptide bond hydrolysis (or aminolysis) is a very efficient process. At saturating concentrations of the carbonyl donor substrate, the reaction can be achieved in 0.01 s or less ($k_{cat} = 100$ s^{-1} or more). The question then arises as to how the enzyme's active site operates.

The scissile $\overset{O}{\overset{\|}{C}}{-}N$ peptide bond has a permanent dipole of 0.72 eÅ that results in a partial positive charge on the carbonyl carbon atom. Rupture of the bond requires the concerted action of an electrophile which polar- izes the C=O bond and a system which is able to perform both nucleophilic

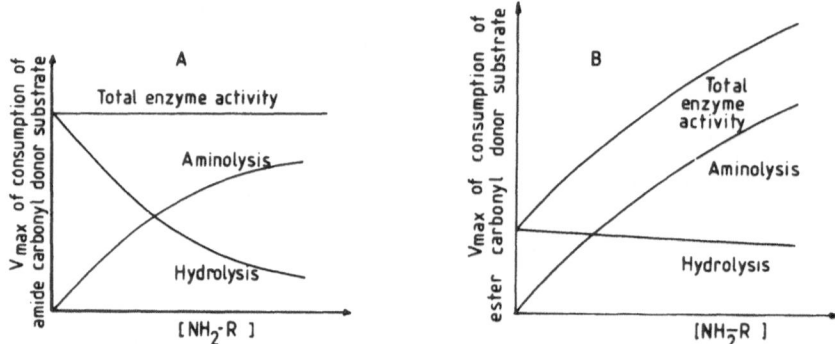

Fig. 8. Effect of an alternate nucleophile ($NH_2{-}R$ = L-alanine amide) on the α-chymotrypsin-catalysed hydrolysis and aminolysis of acetyl·-L-phenylalanine p-trimethylammonium anilide (A) and acetyl-L-phenylalanine methyl ester (B). The rate determining step is acyl enzyme formation in (A) and acyl enzyme breakdown in (B).

attack of the carbonyl atom and proton donation on the nitrogen atom (Figure 9). The same mechanism applies to the rupture of a $\overset{O}{\overset{\|}{C}}$–O ester bond. To all appearances, these roles are played in the enzyme active site by (using the numbering of chymotrysin): a pair of hydrogen bonds from the backbone NH of Ser[195] and Gly[193] (acting as electrophile) and the triad Ser[195] ...His[57] ...Asp[102] (acting as proton abstractor and donator) (Kossiakoff and Spencer, 1981).

The putative mechanism (Figure 10) implies that positioning of a carbonyl donor substrate within the enzyme active site launches a series of reactions which involve: 1) abstraction of the $O\gamma$ proton of Ser[195] by the $N\varepsilon_2$ of His[57] thus activating a nucleophile for attack of the substrate carbonyl; 2) formation of a tetrahedral adduct whose oxyanion hole is stabilized by hydrogen bonding with the NH groups of Ser[195] and Gly[193]; and 3) back delivery of the proton from His[57] to the substrate amino group, causing expulsion of the leaving group P_1 and formation of the acyl enzyme (with restoration of a non-tetrahedral planar trigonal carbonyl carbon). Subsequent enzyme declaration and release of product P_2 would occur through similar steps with this time the exogenous nucleophile HY acting as acceptor. In this mechanism, the carbonyl of Asp[102] remains all the time unprotonated and acts as an orientor of the whole system. Moreover, the presence of water molecules in the enzyme active site is ignored although at least one of them is known to be located between Ser[195] and His[57].

At variance with the above picture, <u>ab initio</u> calculations (Dive et al., 1984 and unpublished data of G. Dive) show that in the trimer methanol ...imidazole...formate or the tetramer methanol...H_2O...imidazole...formate used as models of the Ser...His...Asp triad, an internal proton transfer cannot occur from CH_3OH (i.e. Ser[195]) to imidazole (i.e. His[57]), nor from imidazole (i.e. His[57]) to formate (i.e. the carboxylate head of Asp[102]). Moreover, using the system methanol + formamide as model of the reaction between Ser[195] and a scissile peptide bond, calculations also show that bond rupture occurs only when the following zwitterion limit structure is reached

In the other limit structure

the bond C–N is increased in length but rupture does not occur because of the strong electrostatic interaction between the partial positive charge on the carbonyl carbon and the free pair on the nitrogen atom. Finally, calculations also suggest that one water molecule has to be involved in the mechanism of acyl enzyme formation (Figure 11). By acting as hydrogen donor and acceptor, H_2O may well be the real proton carrier of the enzyme charge relay. Depending on the nature of the substituents R and R' (Figure 11) and the strength of the polarization of the C=O bond, the dihedral angles α and β adopt distinct orientations with regard to the bound substrate, thus influencing the ease of proton transfer. Under optimal conditions, proton transfer and enzyme acylation occur without significant energy barrier.

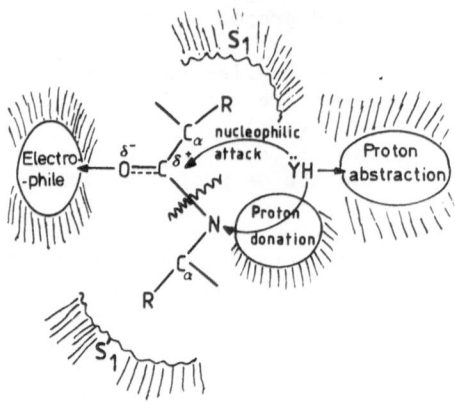

Fig. 9. Rupture of a peptide (amide) bond.
S_1 and S_1' = binding subsites in
the enzyme cavity.

SERINE DD-PEPTIDASE AND SERINE β-LACTAMASE-CATALYSED ACYL TRANSFER REACTIONS
FROM CARBONYL DONOR PEPTIDES AND ESTERS

Reaction 1 developed for chymotryspin applies to the interactions
between the DD-peptidases and their substrates. Figure 12 illustrates
the reaction flux as it occurs when the R61 and R39 serine DD-peptidases
interact with the tripeptide Ac_2-L-Lys-D-Ala-D-Ala used as analogue of a
natural carbonyl donor pentapeptide ∿L-Ala-γ-D-Glu-L-Xaa-D-Ala-D-Ala. The
same reaction pathway occurs with the ester analogue Ac_2-L-Lys-D-Ala-D-
lactate.

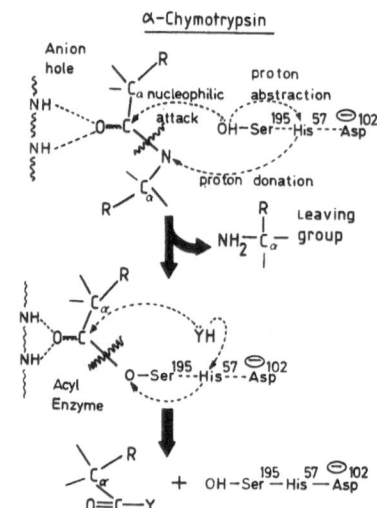

Fig. 10. Putative mechanism of
chymotrypsin-catalysed.
acyl transfer reaction.
HY = exogenous nucleo-
phile; ∿NH∿NH∿= NH
groups of Ser[195] and
GLY[193].

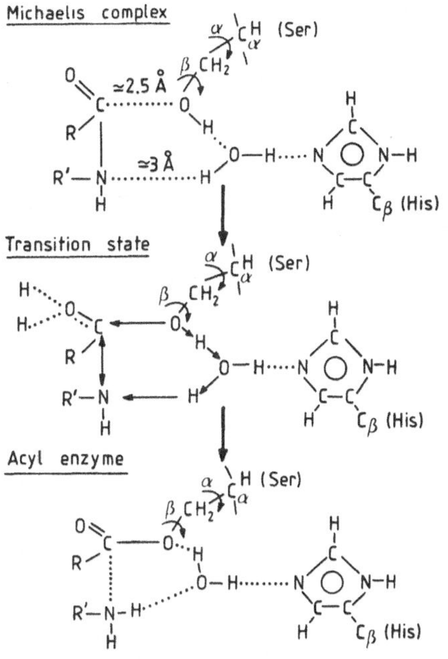

Fig. 11 Concerted reorganization
of the system carbonyl
donor...serine...water...
histidine leading to
peptide bond rupture and
formation of acyl enzyme

The R61 and R39 enzymes are excreted during growth by Streptomyces R61
and Actinomadura R39, respectively. They have been purified to protein
homogeneity and used extensively as models for the study of catalysis of
and inactivation of the serine DD-peptidases. Recent reviews of these
enzymes and list of references can be found in Ghuysen et al., 1979;
Ghuysen et al., 1980; Ghuysen et al., 1981a; Ghuysen et al., 1981b;
Ghuysen et al., 1983; Ghuysen et al., 1984 and Ghuysen, 1984. The R61
enzyme has been crystallized and its three-dimensional structure at 2 Å
resolution is well in progress (Kelly et al., 1985). The active site of
the R61 enzyme includes the sequence Val-Gly-Ser*-Val-Thr-Lys (Figure 6)
(Frère et al., 1976; Kelly et al., 1985; and unpublished data of
B. Joris) and thus has the typical sequence (Gly or Ala)-Ser-Xaa-Xaa-Lys
found in the E. coli DD-peptidases 1A, 1B, 3 and 5. The active-site serine
region of the R61 enzyme also shows close similarity with the active-site
region of the β-lactamases of class C (Figure 6).

As shown in Figure 12, initial recognition of the amide (or ester)
carbonyl donor by the enzyme, binding energy, proper alignment with regard
to the active-site serine residue (and other functional groups) and
catalytic efficacy rely 1) on charge pairing between the terminal carboxylate of the substrate and some cationic group of the enzyme (most likely,
a lysine residue in the case of the R61 enzyme); and 2) on the complementation of the enzyme cavity by the substrate. Enzyme activity strictly
requires a D-Ala at the penultimate position; is decreased but not
abolished when Gly or a D-amino acid other than D-Ala occurs at the carboxy-
terminus (L-Ala at this position abolishes activity); and requires a long
side chain at the L-centre.

Serine   DD-peptidase +   Ac$_2$-L-Lys—D-Ala—D-Ala

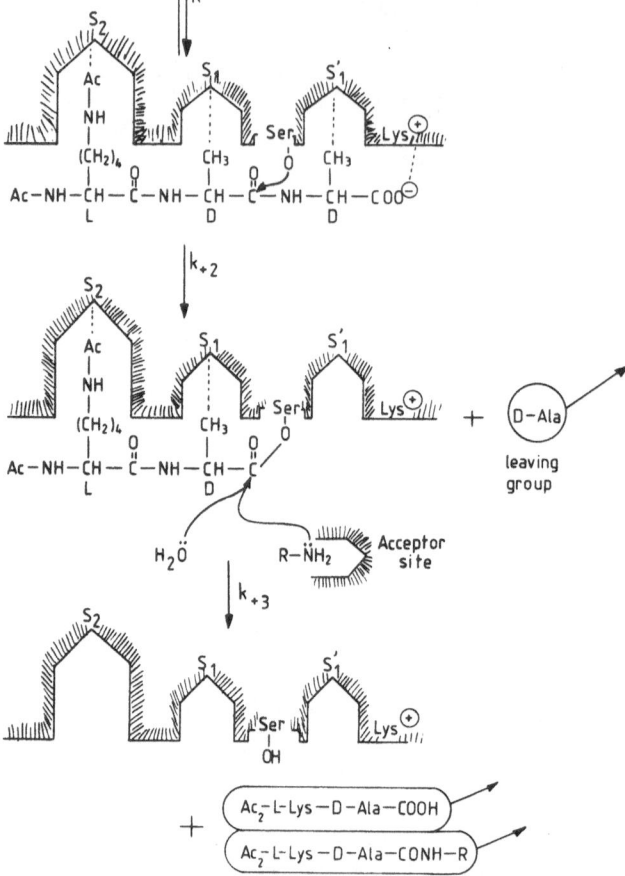

Fig. 12 Schematic view of covalent catalysis by
the serine DD-peptidases. The carbonyl
donor substrate is Ac$_2$-L-Lys-D-Ala-D-Ala.

As shown in Figure 12, the leaving group P$_1$ of enzyme acylation by the
amide substrate is D-Ala (or D-lactate from the ester substrate) and the
reaction product P$_2$ is either the dipeptide Ac$_2$-L-Lys-D-Ala (if the only
exogenous nucleophile present is H$_2$O) or a mixture of dipeptide and transpeptidated peptide Ac$_2$-L-Lys-D-Ala-CONH-R (if the reaction is carried out
in a water solution containing an alternate nuclerophile R-NH$_2$).

The transpeptidated peptide Ac$_2$-L-Lys-D-Ala-CONH-R is not a trivial
reaction product. Indeed, effective chanelling of the enzyme activity to
transpeptidation occurs at millimolar concentrations of a suitable amino
compound (which successfully competes with 55.5 M H$_2$O). Moreover, the R61
and R39 enzymes have highly structured amino acceptor sites which exhibit
distinct specificity profiles and these differences reflect distinct
structural features in the wall peptidoglycans of the relevant producing
strains. Estimation of the K$_{cat}$ values of the catalysed reactions suggests
that the amino acceptor does not bind to the free enzyme, nor to the
Michaelis complex but reacts only as an alternate nucleophile at the level
of the acyl enzyme (as observed with chymotrypsin; see Figure 10). Such
a model probably applies to the DD-peptidases, for example the E. coli

proteins 4, 5 and 6, known to catalyse concomitant carboxypeptidation and transpeptidation reactions.

How does reaction (1) and Figure 12 apply to the DD-peptidases, for example the E. coli proteins 1A, 1B and 3, which are deprived of carboxy-peptidase actitivity and catalyse only transpeptidation reactions? The answer is not known for the E. coli DD-peptidases but it is known for a 26,000-Mr DD-peptidase bound to the plasma membrane of Streptomyces K15.

The K15 DD-peptidase has been purified to protein homogeneity (in the presence of cetyltrimethylammonium bromide) and has been investigated in detail (Nguyen-Distèche et al., 1985). Upon reaction with Ac_2-L-Lys-D-Ala-D-Ala or Ac_2-L-Lys-D-Ala-D-lactate and H_2O as the only available nucleo-phile, acyl enzyme accumulation occurs. However, while the D-lactate which is released from Ac_2-L-Lys-D-Ala-D-Lac has no effect on the hydrolysis of the ester substrate which proceeds until completion, the D-Ala which is released from Ac_2-L-Lys-D-Ala-D-Ala is reutilized as acceptor of a transfer reaction which maintains the amide substrate at a constant level and prevents it from being hydrolyzed. When present in the reaction mixture, a suitable amino acceptor (structurally related to the Streptomyces wall peptidoglycan) suppresses acyl enzyme accumulation, abolishes the acceptor activity of water and successfully competes with the acceptor activity of the D-Ala released from the amide substrate. As a consequence, both amide and ester carbonyl donors are quantitatively transpeptidated until complete consumption, i.e. the K15 DD-peptidase functions as a strict transpeptidase. Strong experimental evidence also suggests that the amino acceptor does not behave as a simple alternate nucleophile but that it influences the initial binding between the enzyme and the carbonyl donor and enhances the efficacy of the ensuing enzyme acylation step.

By analogy with the K15 DD-peptidase, the E.coli proteins 1A, 1B and 3 may lack hydrolysis potency on D-Ala-D-Ala terminated carbonyl donors because the released D-Ala has high acceptor activity. In addition, these DD-peptidases may have low intrinsic catalytic activity unless a suitable amino acceptor (produced by prior or simultaneous transglycosylation of the disaccharide peptide units) enhances the efficacy with which the carbonyl donor acylates the protein.

The Km values for the reactions between the R61, R39 and K15 DD-peptidases and the amide and ester carbonyl donor analogues are relatively high, ranging between 0.1 and 10 mM. Substrate binding is thus relatively weak, a situation which, as discussed in Fersht (1984) favors maximization of the overall reaction rates. With Ac_2-L-Lys-D-Ala-D-Ala, the turnover number values are 55 s^{-1} for the R61 enzyme and 17.5 s^{-1} for the R39 enzyme (hydrolysis pathway) and 0.5 s^{-1} for the K15 enzyme (transpeptidation pathway).

The catalytic pathway of the serine β-lactamases also involves formation of an acyl enzyme. β-lactamases, however, seem to be devoid of peptidase activity. But they catalyse the hydrolysis of certain esters, such as phenylacetylglycylsalicylate, phenylacetylglycyl-D-mandelate and Ac_2-L-Lys-D-Ala-D-Lac (Figure 13) (Pratt et al., 1985). Reaction 1 applies and the presence of alternate nucleophiles causes partitioning of the β-lactamase activity at the level of the acyl enzyme (Knott-Hunziker et al., 1982). The codon for the active site serine of the RTEM β-lactamase has been altered to that for cysteine and the mutant gene has been shown to confer on E. coli host the property to produce a p-chloromercuri-benzoate-sensitive β-lactamase activity (Sigal et al., 1982).

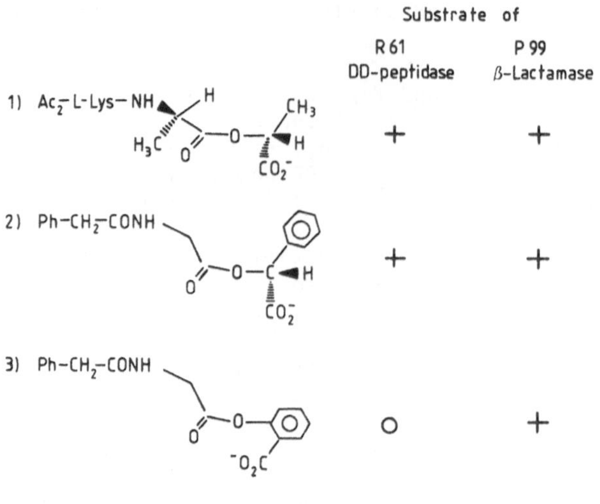

(1) Ac_2-L-Lys-D-Ala-D-Lac ; (2) Phenylacetylglycyl-D-mandelate ;
(3) Phenylacetylglycylsalicylate

Fig. 13. Esters as substrates of the R61 DD-peptidase and P99 β-lactamase.

SERINE DD-PEPTIDASE AND SERINE β-LACTAMASE-CATALYSED ACYL TRANSFER REACTIONS FROM β-LACTAM COMPOUNDS

β-lactam compounds are utilized as carbonyl donors by both serine DD-peptidases and serine β-lactamases. As a consequence of the endocyclic nature of the scissile bond, the leaving group P_1 produced by enzyme acylation remains part of the acyl enzyme and does not diffuse away from the enzyme active site (Figure 14). Consequently, reaction (1) becomes

$$E + D \overset{K}{\rightleftharpoons} E{\cdot}D \xrightarrow{k_{+2}} E\text{-}D^* \xrightarrow[HY]{k_{+3}} E + P \qquad (6)$$

The constant K_{+3} of the interaction between the serine DD-peptidases and the β-lactam compounds has usually a very small absolute value (from $5 \times 10^{-3}\ s^{-1}$ to $1 \times 10^{-7} s^{-1}$) and hence, k_{cat} is negligible. The DD-peptidases become immobilized as acyl enzyme and remain inactivated as long as the acyl enzyme does not break down. The DD-peptidases behave as "penicillin binding proteins" and the β-lactam compounds behave as suicide substrates. A detailed discussion of the phenomenon can be found in Ghuysen et al., 1985. For more details on the interactions between the β-lactams on the R61 and R39 serine DD-peptidases, see Ghuysen et al., 1979; Ghuysen et al., 1980; Ghuysen et al, 1981a; Ghuysen et al., 1981b; Charlier et al., 1983; Ghuysen et al., 1984; Ghuysen, 1984 and Kelly et al., 1985.

For large β-lactam concentrations, not only can the amount of acyl enzyme formed at the steady state be very close to total enzyme but formation of acyl enzyme can be very rapid. For example, for K = 1 mM; $k_{+2} = 10\ s^{-1}$; $k_{+3} = 1 \times 10^{-4} s^{-1}$ and with [β-lactam] = 10 mM, the enzyme catalytic centre "turns over" every 166 min (since $k_{cat} = k_{+3}$), the acyl enzyme at the steady state of the reaction is 99.99% of total enzyme (equation (4), and 99% of the steady state level is reached after 0.5 s (equation (3)).

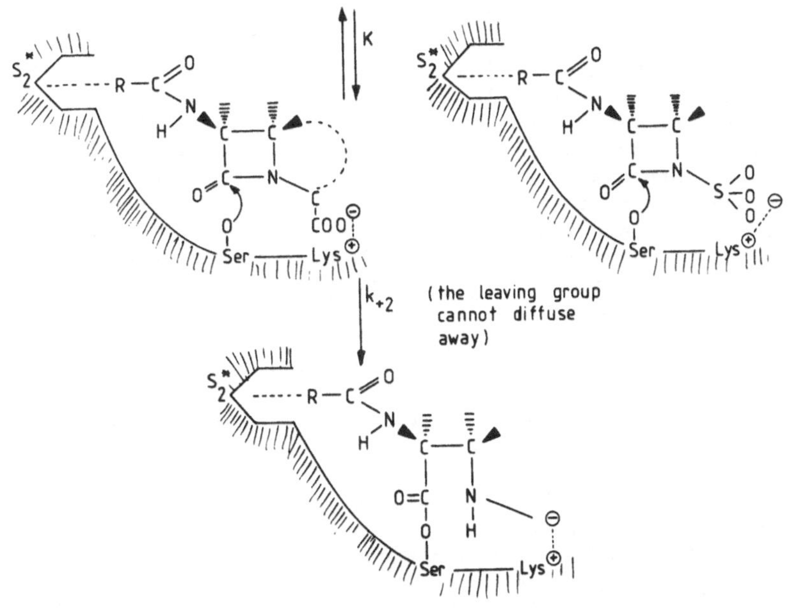

Fig. 14. Schematic view of acyl enzyme formation during
interaction between the serine DD-peptidases and
β-lactam compounds.

Defining the inactivating potency of a β-lactam towards a given DD-peptidase rests upon equation (4). The higher the second-order rate constant k_{+2}/K of enzyme acylation and the smaller the first-order rate constant k_{+3} of acyl enzyme breakdown, the smaller the β-lactam concentration for which $Kk_{+3}/k_{+2}[D]$ becomes negligible, i.e. for which equation (4) becomes $E_o/[E-D*]_{ss} = 1$ so that virtually all the enzyme is immobilized as acyl enzyme at the steady state of the reaction.

The biomolecular rate constant of enzyme acylation k_{+2}/K (and sometimes the individual values of K and k_{+2}) and the first-order rate constant k_{+3} of enzyme deacylation are data experimentally available (Ghuysen et al., 1985). The values of constant K have been determined for the interactions between the R61 and R39 DD-peptidases and several β-lactams. These values range betwen 0.1 and 10 mM (except for the interaction between the R61 enzyme and sulfazecin; K = 32 μM). Hence, as observed with the D-Ala-D-Ala-terminated carbonyl donors, binding of β-lactam compounds is also relatively weak. Both the efficacy of enzyme acylation and the stability of the acyl enzyme are enzyme and β-lactam specific. Thus, for example, the R61 DD-peptidase interacts with 6-aminopenicillanate and benzyl-penicillin with k_{+2}/K values of 0.25 $M^{-1}s^{-1}$ and 13,000 $M^{-1}s^{-1}$, respectively. The corresponding k_{+2}/K values for the interactions with the R39 DD-peptidase are 1,200 $M^{-1}s^{-1}$ and 300,000 $M^{-1}s^{-1}$ (for more details, see Ghuysen et al., 1984). These studies have contributed much to the demonstration that non-planarity of the β-lactam amide bond in the bicyclic β-lactam compounds (and its resulting "supressed amide resonance" or "increased intrinsic reactivity") is not the essential feature responsible for the enzyme inactivating (and antibacterial) potency of these antibiotics.

The constant k_{+3} of the interaction between the serine β-lactamases and the β-lactam compounds has usually very high values and acyl enzyme formation is the rate determining step of the reaction. At saturating concentrations, the reaction flux from good substrates to products may occur in 0.01 s or less. The Km values usually range between 1 and 100 μM. The β-lactamases show widely varying substrate specificities and the interaction is also both enzyme and β-lactam specific. In some cases, relatively long lived acyl enzyme intermediates may be formed. Such a situation occurs for example in the interactions between the RTEM β-lactamase and cefoxitin (k_{+3} = 5 x 10^{-3}s^{-1}) (Fisher et al., 1980) or the Enterobacter cloacae P99 β-lactamase and cloxacillin (k_{+3} = 1.5 x 10^{-4}s^{-1}) (Joris et al., 1985).

The acyl enzyme intermediates formed during reaction of the β-lactam compounds with the serine DD-peptidases and the serine β-lactamases have very different half-lives. However, they have properties in common. One of them is that they break down by acyl transfer to H_2O or other simple nucleophiles such as hydroxylamine. To all appearances, they escape attack by structured amino compounds known to be efficient acceptors in trans-peptidation reactions involving peptide or ester carbonyl donors. These observations suggest that the leaving group (which remains fixed in the enzyme active site) either prevents access of the amino acceptor to or the functioning of the enzyme acceptor site.

A second property is that the acyl enzyme intermediates may undergo rearrangements. Rearrangements destabilize the ester bond in the acyl enzyme formed with the β-lactamases. Thus, the benzylpenicilloyl enzyme formed with various DD-peptidases has been shown to undergo rupture of the C_5-C_6 linkage (Figure 15). The reaction causes release of the leaving group with concomitant formation of phenylacetylglycyl enzyme. The newly formed acyl enzyme then breaks down immediately and the phenylacetylgylcyl moiety can be transferred to both water or a structured amino acceptor. Rupture of the C_5-C_6 linkage is slow and is the rate determining step in acyl enzyme breakdown.

In contrast, the acyl enzyme formed by reaction between β-lactamases and clavulanate, penicillinate sulphone or 6-β-bromo (or iodo) penicillanate undergoes rearrangement into an inactive, stable α-β unsaturated ester E-D^1 (Cartwright and Waley, 1983). The established common feature of these β-lactamase inactivators is that of a branched pathway where β-elimination of the acyl enzyme leads to either an acyl enzyme stable to deacylation or further modification of active site residues

$$E + D \xrightleftharpoons{K} E \cdot D \xrightarrow{k_{+2}} E\text{-}D^* \xrightarrow{k_{+3}} E + P \qquad (7)$$
$$\downarrow{k_{+4}}$$
$$E\text{-}D^i$$

For $k_{+3} \ll k_{+4}$, only inactivation of the β-lactamase occurs. Under other conditions, hydrolysis and rearrangement of the acyl enzyme are competing events and the β-lactamase can be totally inactivated only if $([D]/E)_o$ $(k_{+3} + k_{+4})/k_{+4}$ where $([D]/E)_o$ is the initial ratio of β-lactam concentration to enzyme concentration. Moreover, the final composition of the reaction mixture depends on both the absolute concentration of the β-lactam and the $([D]/E)_o$ ratio value. For $([D]/E)_o < (k_{+3} + k_{+4})/k_{+4}$, the final mixture contains free enzyme, E-D^1 and hydrolyzed β-lactam. For $([D]/E)_o > (k_{+3} + k_{+4})/k_{+4}$, the final mixture contains E-D^1 and both hydrolyzed and intact β-lactam. Figure 16 illustrates the reaction between serine β-lactamase and clavulanate or penam sulphones.

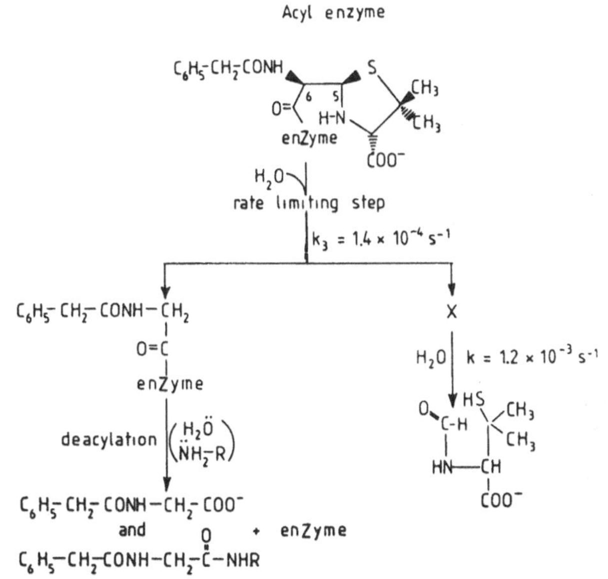

$C_6H_5-CH_2-CONH$

$O=$ $\quad$ 6 $\quad$ 5 $\quad$ S $\quad$ CH_3

$H-N$ $\quad$ CH_3

enZyme

COO^-

H_2O — rate limiting step

$k_3 = 1.4 \times 10^{-4} \ s^{-1}$

$C_6H_5-CH_2-CONH-CH_2$

$O=C$

enZyme

deacylation $\left(\begin{array}{c} H_2\ddot{O} \\ \ddot{N}H_2-R \end{array}\right)$

$C_6H_5-CH_2-CONH-CH_2-COO^-$

and $\qquad$ O $\qquad$ + enZyme

$C_6H_5-CH_2-CONH-CH_2-\overset{O}{C}-NHR$

X

H_2O $\quad$ $k = 1.2 \times 10^{-3} \ s^{-1}$

O $\quad$ HS $\quad CH_3$

$C-H$ $\quad CH_3$

$HN-CH$

COO^-

Fig. 15. Rearrangement of the acyl enzyme formed
between benzylpenicillin and the R61
serine DD-peptidase. Rupture of C_5-C_6
generates phenylacetylglycyl enzyme and
an intermediate X which itself gives
rise to N-formyl-D-penicillamine. Z =
hydroxyl group of the active-site serine.

THE ACTIVE SITE OF THE SERINE DD-PEPTIDASES AND SERINE β-LACTAMASES

Defining the structure of a given enzyme to atomic resolution can be
approached only by crystallography. Binding study of enzyme ligands
(substrates and inactivators) by difference electron density map permits
identification of the active site, and refined data lead to visualization
of the amino acid residues that may play important roles in interaction
energy and catalysis.

Although the first successful crystallization of the RTEM β-lactamase
was reported in 1979 (DeLucia et al., 1980), the three-dimensional structure
at high resolution (1.5 - 2 Å) of any serine β-lactamase is not yet known.
The structure of 5 Å of the S. aureus β-lactamase has been established
(Moult et al., 1985) and, recently, additional β-lactamases have been
crystallized (Dideberg et al., 1985 and unpublished data of O. Dideberg).

Much more information is available with regard to the active site of
at least one serine DD-peptidase. The three-dimensional structure of the
36,000-Mr R61 DD-peptidase at high resolution, the folding of the polypeptide chain and the identification of side chain atoms have been
provisionally established (Figure 17)(Kelly et al., 1985). Although not
yet completed, this work demonstrates that the R61 DD-peptidase has no
structural relatedness with any other known serine peptidase. Diffusion
of a cephalosporin and other β-lactam compounds in the protein crystal
has permitted identification of the active site and visualization of the
active-site serine pointing to the α-face of the bound β-lactam module.
The active-site serine is situated close to the amino-terminus of the
polypeptide. Fournier synthesis of the complex formed with di-isopropyl-
fluorophosphonate (an active-site directed reagent of the serine proteases)
shows only one peak located in the enzyme active site.

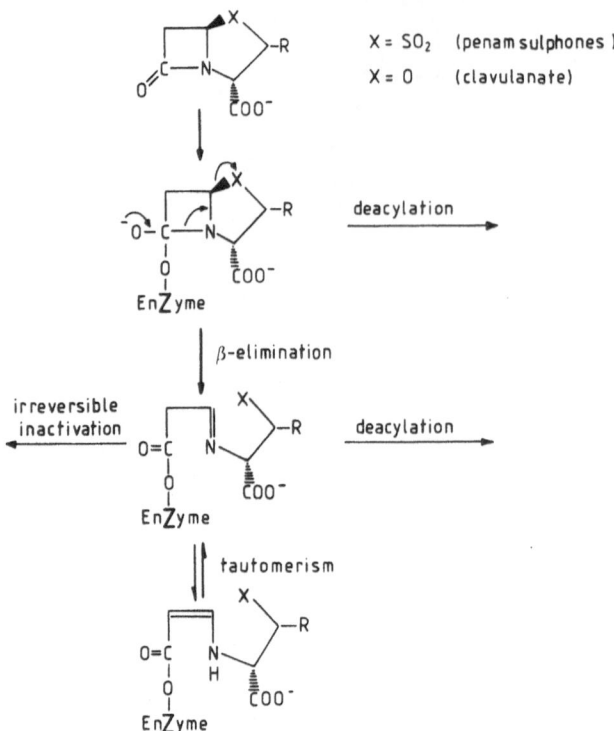

X = SO₂ (penam sulphones)

X = O (clavulanate)

deacylation →

β-elimination

irreversible inactivation ←

deacylation →

tautomerism

Fig. 16. Rearrangement of the acyl enzyme formed
between clavulanate or penam sulphones
and a serine β-lactamase. Z = hydroxyl
group of the active-site serine.

In parallel to these structural works, other studies are on their
way concerning the spatial disposition of the "pharmacophores" which, in
the substrates and inactivators of the DD-peptidases and β-lactamases,
define characteristic patterns that are important features for the
orientation of the whole molecules within the enzyme active site and their

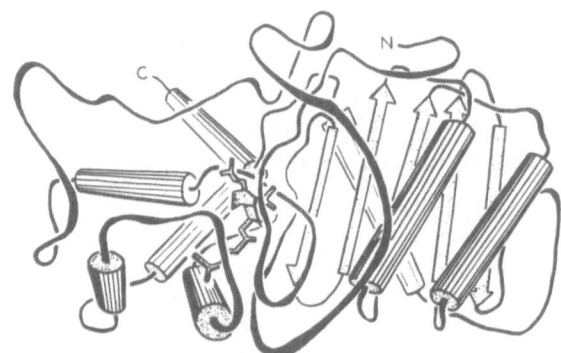

Fig. 17. Folding of the polypeptide chain of
the serine DD-peptidase of Streptomyces R61. Cylinders and arrows
represent α-helices and β-strands,
respectively. Cephalosporin C is
shown in the β-lactam binding site
(Kelly et al., 1985).

reactivity towards the catalytically important side chains. In this respect, the electrostatic potential is probably the most accurate "identity card" of a given molecule in that the electronic distribution, influenced by the conformation, defines the limits of the molecular volume. Calculation of the electrostatic potential of a given molecule (using the CNDO approximation) generates contours of equipotential values and localized electrostatic wells. At each point, the electrostatic potential represents the value, at the first-order of perturbation, of the interaction energy of the molecule with a unitary point charge, i.e. a proton.

The scissile bond in D-Xaa-D-Xaa-terminated peptides and β-lactam compounds is functionally equivalent (Figure 18) in that it is susceptible to attack with varying efficacy by the active-site serine of the DD-peptidases and β-lactamases. Yet, peptides and β-lactam compounds lack overall structural relatedness and the scissile bonds are far from being isosteric to each other.

Calculation of the electrostatic potentials (Lamotte-Brasseur et al., 1984) has shown that all the selected D-Xaa-D-Xaa-terminated and β-lactam conformers have common reactive features around the carbonyl of the scissile

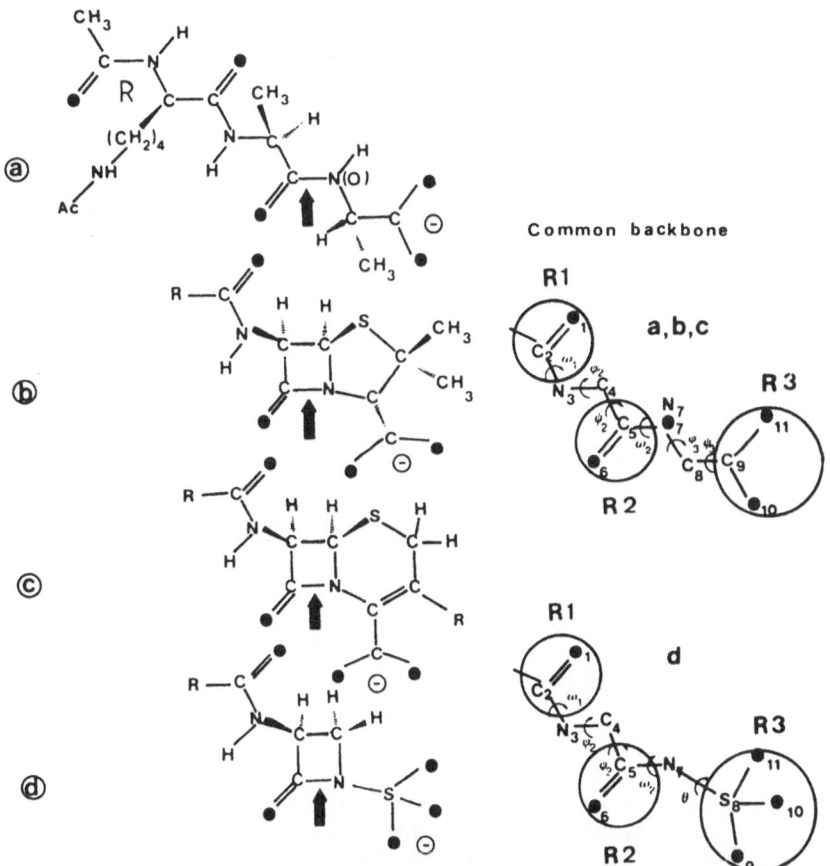

Fig. 18. Structure and common backbone of (a) Ac$_2$-L-Lys-D-Ala-D-Ala (and Ac$_2$-L-Lys-D-Ala-D-Lac); (b) penams; Δ^3-cephems and (4) monobactams. Schematic disposition of the three pharmacophores R$_1$, R$_2$ and R$_3$.

bond (Figure 19). However, depending on the conformation of the peptide backbone, the presence of bulky side chains, the type of bicyclic or monocyclic framework in the β-lactam compounds, the presence of ionized or electron-withdrawing substituents, etc., they show wide variations in the spatial disposition and strength of the positive and negative potentials. As a corollary, these variations must generate enzyme-ligand associations of widely varying complementarity and productiveness.

The characterization of DD-peptidases and β-lactamases by X-ray crystallography and the description of the molecular structures of substrates and inactivators in terms of electrostatic potentials are important advances. They bring the final goal closer, which is a full understanding of the molecular interactions between partners. It remains, however, that calculation of the electrostatic energy of long distance interactions and of the whole interacting systems have to be further developed.

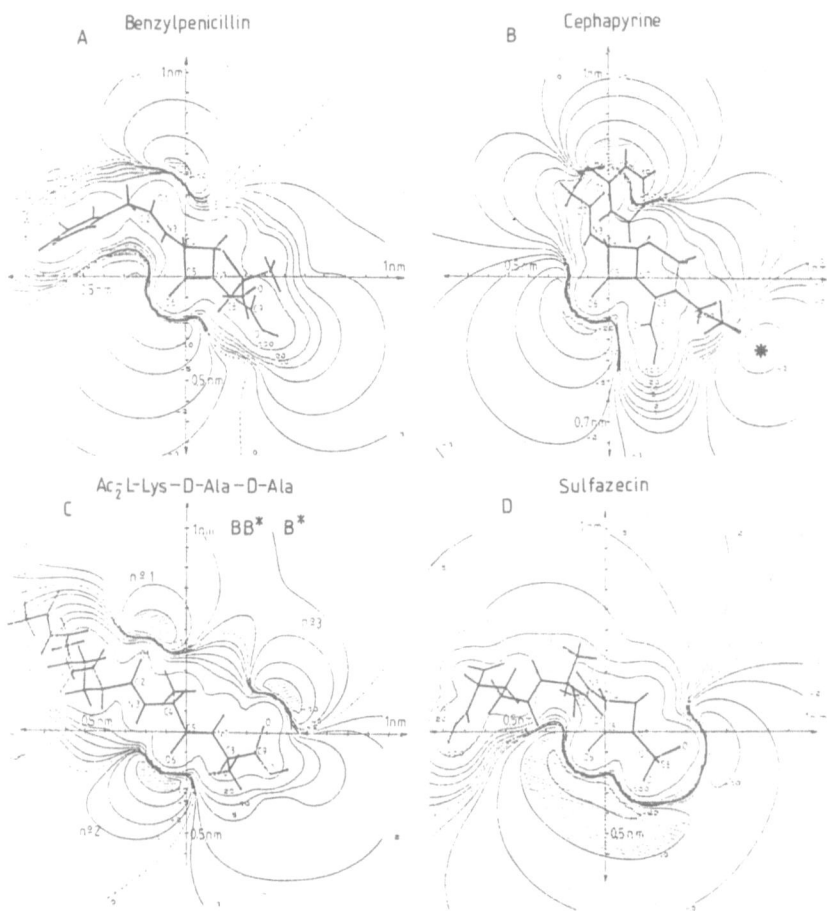

Fig. 19. Potential electrostatic maps of A) benzylpenicillin; B) cephapyrine; C) Ac$_2$-L-Lys-D-Ala-D-Ala (conformer BB*B*) and D) sulfazecin. The maps show contours of equipotential values. Shadowed areas are negative potential wells with contours of - 10 Kcal/mol. The sections of the molecules contain the plane formed by the scissile CONH bond in the peptide and the β-lactam compounds (Lamotte-Brasseur et al., 1984).

Thermolysin and carboxypeptidase A are well-known Zn^{++} peptidases. The putative mechanism (Figure 20) of the catalysed-hydrolysis of peptide or ester bonds by these enzymes is that (using the numbering of thermolysin): i) Glu[143] acts as proton abstractor heightening the nucleophilicity of the zinc-bound water; ii) Zn^{++} acts as an electrophile; and iii) His[231] facilitates proton donation to the nitrogen (oxygen) atom of the scissile bond at the level of the tetrahedral intermediate. Collapse of the intermediate with re-entry of a water molecule causes the release of the reaction products. No covalent intermediate is formed during the process. The Zn^{++} peptidases are strict hydrolases, they do not perform transpeptidation (aminolysis) reactions.

The only known Zn^{++} DD-peptidase, which is excreted by <u>Streptomyces albus</u> G during growth, is a strict carboxypeptidase. The folding of the polypeptide chain (212 amino acid residues; sequence known (Joris et al., 1983)) generates two globular domains connected by a single link and the Zn^{++}-containing active site is an open cleft in the large carboxy-terminal domain (Dideberg, 1982). The amino-terminal domain is an "all α" structure and the carboxy-terminal domain is an "α/β" structure (Figure 21). The Zn^{++} DD-peptidase has no amino acid sequence nor three-dimensional relatedness with any known metalloprotease.

Most likely, the three histidine residues at positions 112, 193 and 196 serve as Zn^{++} protein ligands and the guanidinium side chain of Arg[136] is involved in charge pairing with the Ac_2-L-Lys-D-Ala-D-Ala substrate.

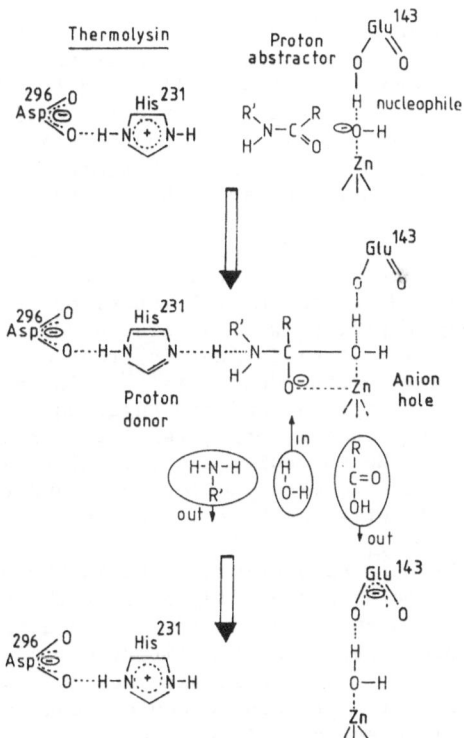

Fig. 20. Putative mechanism of thermolysin-catalysed hydrolysis of peptide bond.

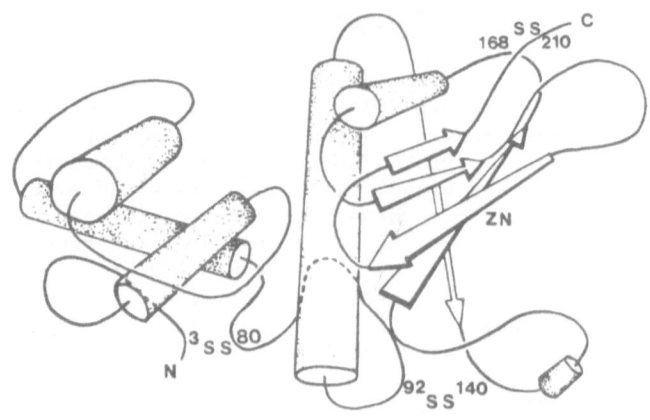

Fig. 21. Folding of the polypeptide chain of the
Zn++ DD-peptidase of _Streptomyces albus_
G. Cylinders and arrows represent α-
helices and β-strands, respectively.
The active site is the open cleft present
in the large domain (Dideberg et al., 1982)

Important catalytic groups would be, in addition to the Zn++ ion acting as electrophile, the imidazole ring of His[190] involved in proton donation and the hydroxyl group of Ser[151], possibly acting as orientor or enhancing the nucleophilicity of a water molecule bound to the Zn++ ion (Charlier et al., 1983).

Productive binding of Ac$_2$-L-Lys-D-Ala-D-Ala to the Zn++ DD-peptidase confers on the peptide a relative orientation of the terminal carboxylate (to Arg[136]) and the carbonyl of the scissile bond (to the Zn++ ion) which falls well outside the limit values found for the corresponding pharmaco-phores in the β-lactam compounds. This situation is consistent with the fact that β-lactam compounds are neither substrates nor effective inacti-vators but, essentially, behave as reversible inhibitors of poor efficacy. Some β-lactam compounds act non-competitively and cause disruption of the protein crystal lattice. Others act competitively, bind to the active site and give rise to isomorphous crystal derivatives. 6-β-Iodopeni-cillinate has been studied in some detail (Charlier et al, 1984). It inactivates the Zn++ DD-peptidase with a first-order rate constant of $7 \times 10^{-4} \text{s}^{-1}$ (limit value). It binds just in front of the Zn++ ion, superimposing the putative proton donor His[190]. In analogy with work carried out with the enkephalinase and the angiotensin converting enzyme, bifunctional compounds possessing both a terminal carboxylate function and, at the other end of the molecule, a thiol, hydroxamate or carboxylate function (for example, racemic mercaptopropionate and the L-isomer of β-mercaptoisobutyrate) inactivate the Zn++ DD-peptidase.

In contrast to the Zn++ DD-peptidase, the 23,000-Mr, Zn++ β-lactamase II of B. cereus has two metal-ion-binding sites (Ambler, 1980). One of them is internally located and might involve three histidine residues of a thiol group (the protein possesses a solitary cysteine residue). The second site would involve a fourth histidine residue and appears to be located on the surface of the protein molecule. Candidates for active site groups do not include serine. A substantial portion of the amino acid sequence is known (R.P. Ambler; private communication). No homology is found with the β-lactamases of class A or C, nor with the Zn++ β-lactamase.

CONCLUSION

Recent advances achieved by gene analysis, establishment of amino acid sequences and mechanistic properties, and elucidation of three-dimensional structures has led to the recognition and characterization of distinct classes of DD-peptidases and β-lactamases. These techniques combined with theoretical chemistry open prospects of designing new types of irreversible inhibitors of these important enzymes, for which there is an urgent need in antibacterial chemotherapy. The serine DD-peptidases and serine β-lactamases utiize the same carbonyl donor substrates but the fate of the substrates differs. Which specific amino acid(s) should be altered in the active site of a DD-peptidase to force the enzyme to work as a β-lactamase and vice-versa? An answer to this question, and to others, can now be approached by protein engineering.

ACKNOWLEDGEMENTS

This work was supported in part by the Fonds de la Recherche Scientifique Médicale, Brussels, Belgium (contract n°3.4507.83), an Action Concertée with the Belgian Government (convention 79/83-I1) and a Convention tripartite between the Région wallonne, Continental Pharma and the University of Liège.

REFERENCES

Ambler, R.P., 1980, Phil. Trans. R. Soc., 289:321-333.

Bergström, S., Ollson, O. and Normark, S., 1982, J. Bact., 150:528-534.

Bicknell, R., Emanuel, E.L., Gagnon, J. and Waley, S.G., 1985, Biochem J., in press.

Broome-Smith, J.K., Edelman, A. and Spratt, B.G., 1983, in: "The Target of Penicillin", J.V. Höltje, R. Hackenbeck and H. Labischinski, eds., Walter de Gruyter, Berlin, pp 403-408.

Broome-Smith, J.K., Edelman, A., Yousif, S. and Spratt, B.G., 1985, Eur. J. Biochem., 147:437-446.

Cartwright, S.J. and Waley, S.G., 1983, Med. Res. Reviews, 3:341-382.

Charlier, P., Dideberg, O., Dive, G., Dusart, J., Frère, J.M., Ghuysen, J.M., Joris, B., Lamotte-Brasseur, J., Leyh-Bouille, M. and Nguyen-Distèche, M., 1983, in: "The Target of Penicillin. The Murein Sacculus of Bacterial Cell Walls. Architecture and Growth", R. Hakenbeck, J.V. Höltje and H. Labischinski, eds., Walter de Gruyter, Berlin, pp 369-386.

Charlier, P., Dideberg, O., Jamoulle, J.C., Frère, J.M., Ghuysen, J.M., Dive, G. and Lamotte-Brasseur, J., 1984, Biochem. J., 219:763-772.

Dalbadie-McFarland, G., Cohen, L.N., Rigth, A.D., Morin, C., Itakura, K. and Richards, J.H., 1982, Proc. Natl. Acad. Sci. USA, 79:6409-6413.

DeLucia, M.L., Kelly, J.A., Mangion, M.M., Moews, P.C. and Knox, J.R., 1980, Phil. Trans. R. Soc. London B., 289:374-376.

Dideberg, O., Charlier, P., Dive, G., Joris, B., Frère, J.M. and Ghuysen, J.M., 1982, Nature, 299:469-470.

Dideberg, O., Libert, M., Frère, J.M., Charlier, P., Zhao, H. and Knox, J.R., 1985, J. Mol. Biol., 181:145-146.

Dive, G., Peeters, D., Leroy, G. and Ghuysen, J.M., 1984, J. Mol. Struc., 107:117-126.

Duez, G., Joris, B., Frère, J.M., Ghuysen, J.M. and Van Beeumen, J., 1981, Biochem. J., 193:83-86.

Fersht, A., 1984, "Enzyme Structure", second edition, W.H. Freeman.

Fischer, J., Belasco, J., Khosla, S. and Knowles, J.R., 1980, Biochemistry, 19:2895-2901.

Frère, J.M., Duez, C., Ghuysen, J.M. and Vandekerckhove, J., 1976, FEBS Lett., 70:257-260.

Ghuysen, J.M., 1968, Bact. Rev., 32:425-464.

Ghuysen, J.M., 1977, in: "The Synthesis, Assembly and Turnover of Cell Surface Components", Cell Surface Reviews, vol. 4, G. Poste and G.L. Nicolson, eds., North-Holland Publishing Co., pp 463-595.

Ghuysen, J.M., 1984, in: "IUPHAR 9th International Congress of Pharmacology, London 1984", J.F. Mitchell and P. Turner, publ., The Macmillan Press Ltd., pp 115-123.

Ghuysen, J.M., Frère, J.M., Leyh-Bouille, M., Coyette, J., Dusart, J. and Nguyen-Distèche, M., 1979, Ann. Rev. Biochem., 48:73-101.

Ghuysen, J.M., Frère, J.M., Leyh-Bouille, M. and Dideberg, O., 1981a, in: "Molecular Basis of Drug Action. Developments in Biochemistry", vol. 19, Th. P. Singer and R.N. Ondarza, eds., Elsevier/North-Holland, pp 3-30.

Ghuysen, J.M., Frère, J.M., Leyh-Bouille, M., Dideberg, O., Lamotte-Brasseur, J., Perkins, H.R. and De Coen, J.L., 1981b, in: "Topics in Molecular Pharmacology", A.S.V. Burgen and G.C.K. Roberts, eds., Elsevier/North-Holland Biomedical Press, pp 63-97.

Ghuysen, J.M., Frère, J.M., Leyh-Bouille, M., Nguyen-Distèche, M. and Coyette, J., 1985, J.J. Bact., submitted.

Ghuysen, J.M., Frère, J.M., Leyh-Bouille, M., Nguyen-Distèche, M., Coyette, J., Dusart, J., Joris, B., Duez, C., Dideberg, O., Charlier, P., Dive, G. and Lamotte-Brasseur, J., 1984, Scand. J. Infectious Diseases, suppl. 42:17-37.

Ghuysen, J.M., Frère, J.M., Leyh-Bouille, M., Perkins, H.R. and Nieto, M., 1980, Phil. Trans. R. Soc. London B., 289:285-301.

Hedge, P.J. and Spratt, B.G., 1984, FEBS Lett., 176:179-184.

Ishino, F. and Matsuhashi, M., 1981, Biochem. Biophys. Res. Commun., 101:905-911.

Ishino, F., Mitsui, S., Tamaki, S., Matsuhashi, M., 1980, Biochem. Biophys. Res. Commun., 97:287-293.

Jaurin, B. and Grundström, T., 1981, Proc. Natl. Acad. Sci. USA, 78:4897-4901.

Joris, B., De Meester, F., Galeni, M., Reckinger, G., Coyette, J., Frère, J.M. and Van Beeumen, J., 1985, Biochem. J., in press.

Joris, B., Dusart, J., Frère, J.M., Van Beeumen, J., Emanuel, E.L., Petursson, S., Gagnon, J. and Waley, S.G., 1984, Biochem. J., 223:271-274.

Joris, B., Van Beeumen, J., Casagrande, F., Gerday, Ch., Frère, J.M. and Ghuysen, J.M., 1983, Eur. J. Biochem., 130:53-69.

Keck, W., Glauner, B. and Schwarz, U., 1985, Proc. Natl. Acad. Sci. USA, in press.

Kelly, J.A., Knox, J.R., Moews, P.C., Hite, G.J., Bartolone, J.B., Zhao, H., Joris, B., Frère, J.M. and Ghuysen, J.M., 1985, J. Biol. Chem., in press.

Kossiakoff, A.A. and Spencer, S.A., 1981, Biochemistry, 20:6462-6474.

Knott-Hunziker, V., Petursson, S., Jayatilake, G.S., Waley, S.G., Jaurin, B. and Grundström, T., 1982a, Biochem. J., 201:621-627.

Knott-Hunziker, V., Petursson, S., Waley, S.G, Jaurin, B., and Grundström, T., 1982b, Biochem. J., 207:315-322.

Lamotte-Brasseur, J., Dive, G., and Ghuysen, J.M. 1984, Eur. J. Med. Chem., 19:319-330.

Moult, J., Sawyer, L., Herzberg, O., Jones, C.L., Coulson, A.F.W., Green, D.W., Harding, M.M. and Ambler, R.P., 1985, Biochem. J., 225:167-176.

Nakagawa, J.I., Tamaki, S. and Matsuhashi, M., 1979, Agric. Biol. Chem., 1379-1380.

Nakamura, M., Maruyama, I.N., Soma, N., Kato, J.I., Suzuki, H. and Hirota, Y., 1983, Mol. Gen. Genet., 191:1-9.

Nguyen-Distèche, M., Leyh-Bouille, M., Pirlot, S., Frère, J.M. and Ghuysen J.M., 1985, J. Bact., submitted.

Pratt, R.F., Faraci, W.S. and Gowardhan, C.P., 1985, Analytical Chem.,
    in press.
Richmond, M.H., Bennet, P.M. Choi, C.L., Brown, N., Brunton, J., Gunsted,
    J. and Wallace, L., 1980, Phil. Trans. R. Soc. London B., 289:349-359.
Sigal, I.S., Harwood, B.G. and Arentzen, R., 1982, Proc. Natl. Acad. Sci.
    USA, 79:7157-7160.
Spratt, B.G., 1983, J. Gen. Microbiol., 129:1247-1260.
Suzuki, H., Van Heijenoort, Y., Tamura, T., Mizoguchi, J., Herato, Y. and
    Van Heijenoort, J., 1980, J. FEBS Lett., 110:245-249.

THE INDUSTRIAL PRODUCTION OF MONOCLONAL ANTIBODIES IN CELL CULTURE

J.R. Birch, K. Lambert, R. Boraston, P.W. Thompson,
S. Garland and A.C. Kenney

Celltech Ltd., 244-250 Bath Road, Slough SL1 4DY
Berks, U.K.

INTRODUCTION

The development of the hybridoma technique by Kohler and Milstein in 1975 made it possible to produce for the first time monoclonal antibodies of constant and defined specificity in potentially unlimited quantities. Monoclonal antibodies have become important tools in the research laboratory and have attracted widespread commercial interest. In the field of diagnostics there has been a rapid development of assay systems based on monoclonal antibodies for the detection and measurement of, for instance, drugs, hormones, infectious diseases and blood group antigens (Nowinski et al., 1984; Voak and Lennox, 1983). The feasibility of using suitably labelled antibodies to diagnostically "image" tumor tissues in vivo is being actively investigated by a number of groups (see for example Mach et al., 1981). It is recognised that if monoclonal antibodies can be used to detect tumor tissue in this way then they may also be useful in therapy, for instance, as a means of delivering cytotoxic agents specifically to cancer cells (Marx, 1982; Vitella, 1982; Miller, 1982). Another exciting application which exploits the extreme specificity of the antibody-antigen interaction is immunopurification. This method is particularly useful for the purification of high value products such as interferon (Secher and Burke, 1980) which may be present in dilute solution in crude mixtures. We anticipate this technique becoming more widely applied as antibodies become more readily available and less expensive (Hill et al., 1985). The increased application of monoclonal antibodies has resulted in a need to produce large quantities (grams to kilograms per year) at appropriate cost. This has led our group and others to develop new production methods. Our own developments in fact arose from the need to produce monoclonal ABO blood typing reagents (Voak and Lennox, 1983) and these were the first bulk monoclonal products to reach the market.

MONOCLONAL ANTIBODY PRODUCTION METHODS

Broadly we can distinguish between two types of production system. Hybridoma cells are grown either in vivo as ascites tumors in mice and rats or in vitro using tissue culture techniques. The ascites method has been the route most commonly used to produce small (gram) quantities of antibodies but is favoured by some for large scale production. Although this procedure produces a highly concentrated product, the yield per animal is small (typically 50mg/mouse) and in consequence scale up requires the use of enormous numbers of animals. To produce one kilogram of antibody may require 20,000 mice. The alternative approach which we strongly favour is to use in vitro cell culture. This method has already been used for hybridoma culture but typically on a small scale in static flasks or roller bottles. The in vitro route has a number of distinct advantages:

- The potential exists for scaling up the size of the culture vessel allowing one to benefit from the economies of scale. Large quantities of antibody can be made at low cost.

- The process can be engineered to be highly reproducible.

- There is a reduced risk of product contamination by adventitious agents of rodents.

- The presence of extraneous antibody is minimised or avoided.

- Human antibodies can be produced. This is difficult in rodents.

A number of techniques have been proposed for the large scale production of monoclonal antibodies and these are summarised in Table 1.

HOMOGENEOUS SUSPENSION CULTURE SYSTEMS

Hybridoma cells can be grown in free suspension and can therefore in principle be grown in the types of deep tank fermenter used for micro-organisms. Indeed stirred tank reactors of up to 3000 litres capacity have already been used for the production of foot-and-mouth disease virus vaccine from baby hamster kidney cells and more recently human lymphoblastoid interferon has been produced from Namalva lymphoblastoid cells in an 8000 litre stirred tank fermenter (Pullen et al., 1984). However, relatively little attention has been paid to the production of monoclonal antibodies by such techniques.

In choosing a scale-up route we had to consider the following important criteria which we believe are best achieved in a homogeneous suspension culture system.

Table 1. Cell Culture Methods

Method	Reference
Homongeneous suspension culture	
a. Airlift reactor	Birch et al., 1984
b. Continuous stirred reactor	Fazekas de St. Groth, 1983
	Boraston et al., 1983a, b.
Immobilised or entrapped cell cultures	
a. Hollow fibre perfusion	Wiemann et al., 1983
b. Microcapsules	Geyer et al., 1984
c. Agarose microbeads	Nilsson et al., 1983
d. Ceramic matrix	Marcipar et al., 1983
e. Ceramic cartridge	Lydersen et al., 1985

Aseptic Operation

Anyone familiar with animal cell culture will be aware of the problem of maintaining sterility. We need a system which can be effectively sterilised prior to introducing cells and from which microorganisms can be excluded by appropriate barriers during prolonged operating periods which may extend from one or two weeks for a batch culture up to several months for a continuous culture. While small vessels may be sterilised in an autoclave, larger vessers can only be sterilised in situ. This is done most effectively by steaming at high temperature and pressure and places heavy reliance on good process engineering. This implies for instance the use of hygienic pressure vessels, seal-less pumps and valves and preferably all welded pipework. It is important to eliminate crevices and dead-legs.

Environmental Conditions For Growth

It is essential to provide a system with adequate mixing and heat and mass transfer characteristics to maintain the correct temperature, pH and dissolved oxygen concentration in the culture. However, mixing has to be compatible with the relatively fragile cells with which we are dealing and potentially damaging shear forces must be kept to a minimum.

Process Automation

A manually-operated process plant becomes more difficult to operate and increasingly labour intensive as the scale increases. Hence automation becomes important at a large scale where the application of computer control offers automatic digital sequencing of valves and pumps; analogue control, alarm monitoring and provision of emergency action in case of failures. This can have a considerable impact on costs and security. To take advantage of this facility it is important to devise a process which is amenable to automation; many laboratory scale processes are not.

Ease of product recovery

Monoclonal antibodies are secreted by the hybridoma cells and we require a straightforward means of recovering the antibody. Cell separation is readily achieved by centrifugation or filtration and ultrafiltration techniques allow rapid product concentration. Subsequent purification techniques are likely to be similar whatever the cell culture route used and are considered in more detail below.

Economy of Scale

Scale-up means simply to increase output of the production process and may be achieved in various ways which may or may not provide advantages in terms of cost of the final product. If scale-up consists of simply increasing the number of production units (whether a mouse or a fermenter) without increasing unit size then both capital and labour costs will increase linearly with increase in scale. This provides no cost advantage. However, if scale-up is addressed by increasing unit size rather than unit number, significant cost advantages can be realised because capital cost of equipment increases according to the 0.6 power ratio of capacity and labour costs effectively remain static. As the scale increases consumable costs such as culture media form a greater proportion of the total production costs and increased emphasis then needs to be placed on reducing these costs.

The particular type of fermenter which we have used for homogeneous suspension culture is the airlift reactor. The growth of animal cells in this type of vessel was first described by Katinger et al. (1979) who considered it to have particularly appropriate characteristics for shear sensitive cells. The airlift principle is illustrated in Figure 1.

The vessels are fabricated of glass at the laboratory scale (5 1) and stainless steel at production scale (100 and 1000 1). The vessels have an internal concentric draught tube. Gas mixtures are introduced into the culture from a sparge ring inside the base of this draught tube. This establishes circulation of the culture because gas hold up causes a density differential between the contents of the draught tube and the outer zone of the vessel. By varying the flow rate and composition of the gas mixture the dissolved oxygen tension and the pH of the culture can be controlled.

This type of fermenter has a number of advantages, one of which is simplicity. The airlift reactor has no moving parts such as the motors and stirrer shafts found on turbine agitated vessels. The absence of shaft seals in turn eliminates a potential route for ingress of contaminants. Suitably low levels of shear stress (as far as can be determined) are realised while maintaining sufficiently high recirculation rates to achieve good mixing of additives, sufficient homogeneity of the culture and prevention of cell sedimentation.

The airlift reactor also has very good oxygen transfer characteristics, which meets the measured oxygen demand of hybridoma cells (Boraston et al., 1983). Although their oxygen requirement is small in comparison with microbial cultures we have found that hybridoma cells are obligate aerobes and there is increasing awareness that many cell culture systems are unable to supply sufficient oxygen to support cell growth at high cell population densities (Fleischaker and Sinskey, 1981; Glacken et al., 1983).

In many small scale culture systems cells are grown in turbine agitated "spinner" cultures with oxygen supplied by simple diffusion from the air in the headspace of the fermenter. In such a system oxygen limitation is frequently encountered (Boraston et al., 1983a). The oxygen transfer rate may be increased by raising the impeller speed but this may lead to unacceptably high shear forces around the impeller. Fazekas de St. Groth

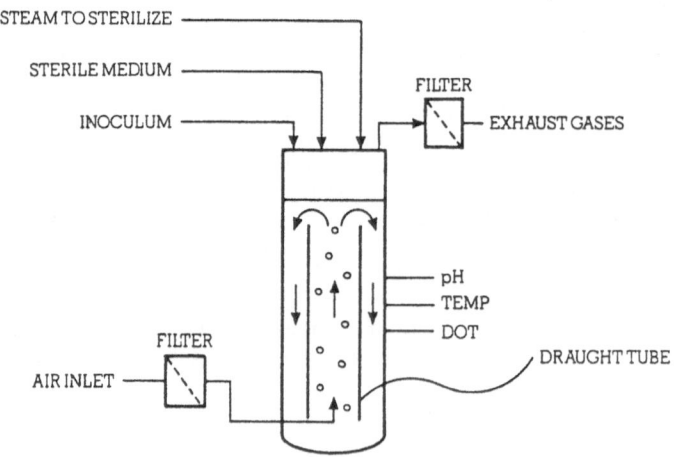

Figure 1. Principle of an Airlift Reactor

found that the impeller speed had to be kept below 60 rpm (equivalent to a tip speed of 0.19 m/s) to prevent damage to hybridoma cells growing in stirred culture in serum free medium. As the scale increases, more effective oxygen transfer systems, such as air sparging are required. The airlift fermenter provides a particularly efficient oxygen transfer system. In Figure 2 we show the oxygen transfer characteristics of 30 to 100 litre airlift fermenters as a function of volumetric air throughput. The arrow indicates the oxygen demand rate of 2×10^6 cells/ml, a typical population density for hybridoma cells approaching stationary phase in batch culture. It will be seen that we can operate well within the oxygen transfer capacity of the fermenter. The slightly reduced oxygen transfer rates at the 100 litre scale reflect the lower aspect ratio of this vessel (10/1) compared with the 30 litre vessel (12/1). The culture recirculation rates in these airlift vessels are typically in the range 0.1 to 0.2 m/s.

To date we have operated airlift fermenters of 5, 10, 30, 100 and 1000 litres working volume. Twenty five different cell lines of mouse, rat and human origin have been grown in these fermenters, all producing monoclonal antibodies. Scale up has been predictable and growth kinetics and antibody synthesis have proved to be similar at all scales examined.

With the 1000 litre vessel we now have a capacity of 3 to 5 kilograms per year and we are designing a 10,000 litre vessel to meet projected future needs.

In addition to the process engineering aspects which we have discussed we have studied in detail the physiology of antibody production. By systematic process optimisation and culture medium design we have been able to increase the antibody yield for many hybriodomas by four to five fold compared with static flasks or roller bottles. We find in common with the published literature (for instance Goding, 1980) that rodent cell lines typically produce between 10 and 100 milligrams of antibody per litre of culture fluid in static flasks. In our fermenters we see a range from about 40 to 500 milligrams per litre with an average yield from cell lines grown

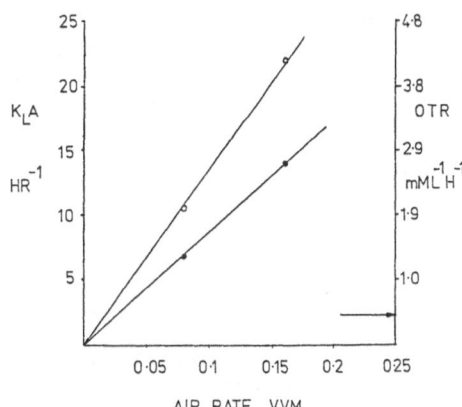

Fig. 2. Effect of air rate on oxygen
        transfer rate in airlift
        fermenters, measured by a
        dynamic method using an oxygen
        electrode. Arrow indicates
        oxygen consumption rate of
        2×10^6 cells/ml.
        ● - ● 100 litre vessel
        o - o 30 litre vessel

to date of 102 milligrams per litre. We believe that there is still considerable opportunity for further process optimisation but we also recognise the importance of understanding the basis of the inherent variation in antibody synthesis between cell lines and devising methods to generate more productive cell lines.

Figure 3 shows growth and antibody production by a mouse hybridoma producing an IgG antibody in the 1000 litre production airlift vessel, a photograph of which is shown in Figure 4. This fermenter incorporates a high degree of automation with a computer controlling most valve sequences, in sterilisation for instance, in addition to monitoring, controlling and alarming key functions such as dissolved oxygen tension, pH and temperature. In the experiment shown in Figure 3, the hybridoma cells grew exponentially with a pollution doubling time of 20 hours to a maximum population density of 2×10^6 cells/ml. Synthesis of antibody occurred during the growth phase and, interestingly, continued during the decline phase. Indeed, 60% of the total antibody is synthesised after maximum population density has been reached.

Although most of our work has been carried out with rodent cell lines, we have successfully grown human antibody producing cell lines in airlift culture. This is important because monoclonal antibodies of human origin are likely to become increasingly important in therapeutic applications. Figure 5 shows growth and antibody production by a human EBV transformed lymphoblastoid cell line. In general human lines are regarded as poor producers of antibody but it will be seen that at least for this cell line, yields comparable with rodent cell lines can be achieved.

CONTINUOUS CULTURE

The data shown in Figures 3 and 5 follow the typical profiles for batch culture of microorganisms. A brief lag phase is followed by exponential growth and then stationary and decline phases. Under batch culture conditions the environment is constantly changing - nutrients are depleted while metabolites and product accumulate. In such a system it is difficult to study the parameters that control cell metabolism and product synthesis and this presents something of an obstacle to devising ways of specifically improving product yields. An alternative method, and one which is more

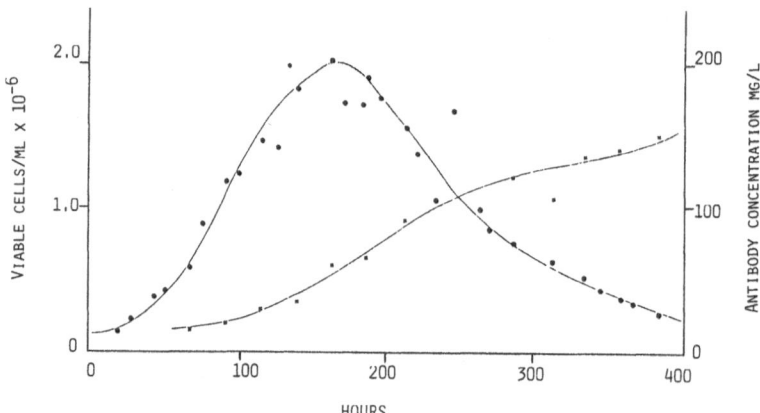

Figure 3. Batch culture of a mouse hybridoma cell line in a 1000 l fermenter.
Viable cells/ml x 10   (—●—)
Antibody concentration (—x—)

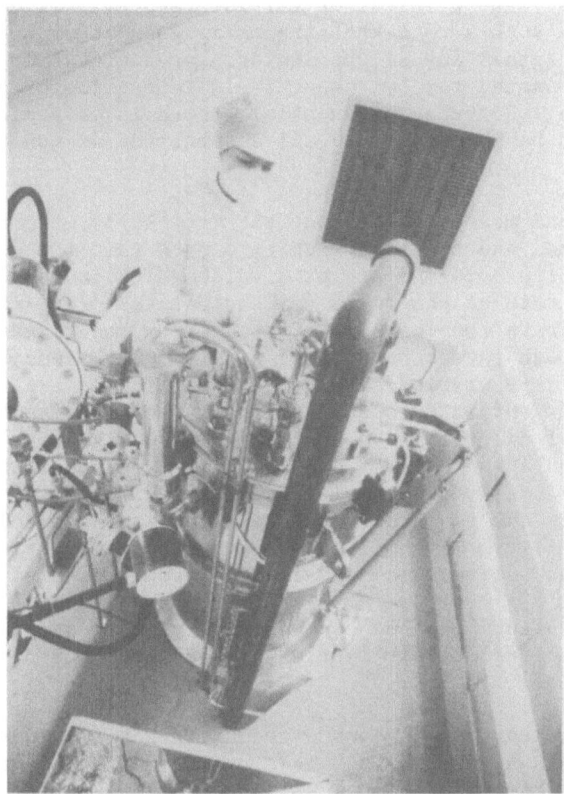

Figure 4.   1000 litre airlift vessel.
Photograph courtesy of
Celltech Ltd.

amenable to close analytical examination, is continuous culture. Continuous
culture differs from batch culture in that a continuous feed of fresh medium
is supplied to the culture and the culture fluid containing cells and cell
products is removed at the same rate.

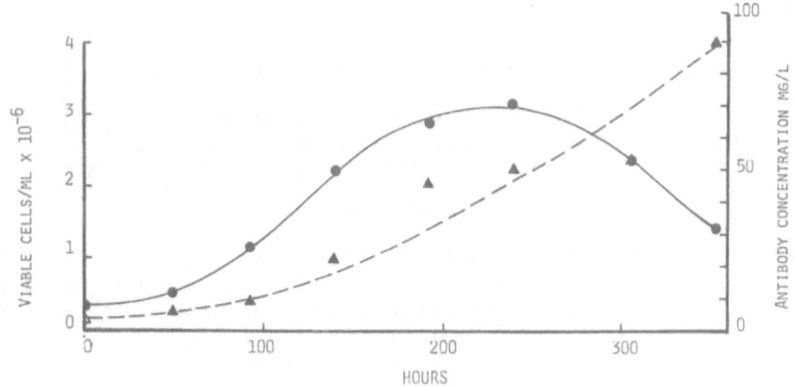

Figure 5.   Batch culture of an EBV transformed human lympho-
blastoid cell line in an airlift fermenter.
Viable cells/ml x 10^{-6} ————•———— ;
Antibody concentration mg/l ----▲----.

In the first, the cytostat or turbidostat (Fazekas de St. Groth, 1983) medium is fed at such a rate that the cells grow at or about their maximum specific growth rate. Any slight increase or decrease in cell population density is compensated for by respectively increasing or decreasing the medium flow rate. Constant exponential growth is maintained and all nutrients should be in excess and all catabolites at non-toxic concentrations.

In the second mode, the chemostat (Pirt, 1975), the medium feed is kept at a predetermined and constant flow rate such that the dilution rate (flow rate per unit culture volume) is at a value which is less than the maximum specific growth rate of the cells (if the dilution rate exceeds the maximum specific growth rate then washout of cells from the fermenter will occur). The specific growth rate of the cells will be determined by the dilution rate and the cell population density will be limited either by depletion of an essential nutrient or by accumulation of a catabolite to a toxic level. Thus by choice of dilution rate and by medium design both the growth rate and the cell population density can be deliberately and independently controlled. Eventually a physiological and environmental steady state is attained in which any measured parameter, for instance cell population density and viability, metabolite and product concentrations, will be constant with respect to time. Again cell growth is truly exponential.

Chemostat culture has been extensively used to study microbial growth and physiology but relatively little work has been performed using animal cells (Tovey, 1980).

In our laboratory we have used the chemostat to study cell growth and antibody synthesis by NBI cells. This hybridoma cell synthesises an IgM antibody against human red blood cell group B antigen and is used as a blood typing reagent (Sacks and Lennox, 1981). A typical steady state for NBI cells growing in oxygen limited chemostat culture is shown in Figure 6 and it can be seen that the measured parameters all remain constant. This constant environment allows much easier analysis of the effect of deliberately altering individual environmental conditions on product synthesis rate, metabolic quotients, etc.

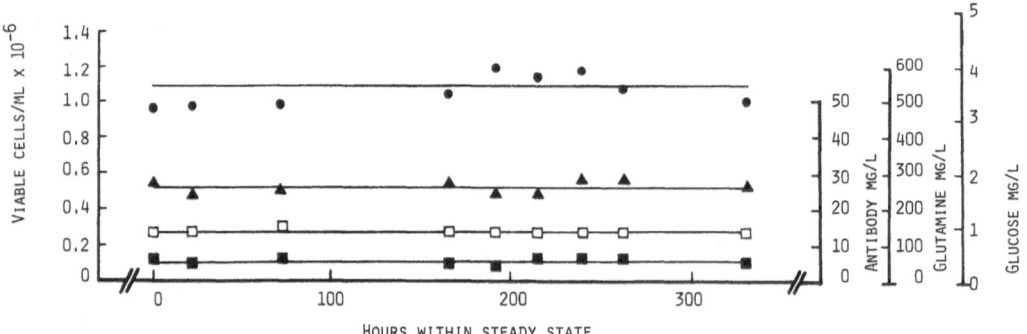

HOURS WITHIN STEADY STATE

Figure 6.   A typical steady state for the mouse hybridoma cell line, NB1, growing in chemostat culture. Oxygen was the growth limiting nutrient and the specific growth rate was 0.021 H^{-1}. The medium feed contained glucose at 4.5 mg/1 and glutamine at 584 mg/1.
Viable cells/ml x 10^{-6} ————•———— ;
Antibody concentration mg/1 ————▲———— ;
Glucose concentration g/1 ————□———— ;
Glutamine concentration mg/1 ————■———— .
(Reproduced from Birch et al., 1984, Academic Press, in press)

We have been surprised to find that antibody synthesis in chemostat culture is stable over prolonged culture periods and under a variety of nutrient limitations (Birch et al., 1984). One might have expected this type of culture system to have exerted a selective pressure against cells synthesising antibody which one presumes is non-essential and does not contribute to the generation of new biomass.

Apart from the advantages of the chemostat already noted, viz: a constant environment and the ability to control growth rate and the nature of the limiting nutrient, this type of culture system may also be useful as a production technique. In particular, chemostat culture may offer more efficient use of fermenter capacity than a batch system. Once established a continuous culture can be maintained at high cell density and high product output for many months. Furthermore, the additional parameters which one can control (growth rate and nutrient limitation) allows one greater scope for optimization of biomass and product formation than can be achieved in batch culture. At the laboratory scale, Fazekas de St. Groth (1983) has grown ten different hybridoma cell lines in continuous culture. We have operated chemostats at the laboratory scale (1 to 5 litres) for periods up to 6 months and we have now modified a 30 litre airlift vessel to operate continuously.

Continuous Culture with Cell Feedback

The productivity of the chemostat can be increased if instead of allowing cells to leave the vessel in the effluent stream, they are retained in the vessel. The mathematics underlying this principle are explained by Pirt (1975). We have modified a 5 litre airlift vessel to give complete retention of viable cells. The consequence of this is that one can achieve increased population densities and use increased dilution rates. Indeed one can theoretically increase the dilution rate beyond the maximum specific growth rate of the cell population without washout. Hence the overall product output rate can be increased compared with a normal chemostat. This is illustrated in Table 2. NB1 cells were grown either as a conventional chemostat (D = 0.02) or as a chemostat with feedback (D = 0.04). The growth limiting nutrient in this experiment was glutamine. Virtually no viable cells were removed in the effluent stream when the fermenter was operating with cell feedback. However, the effluent did contain dead cells and cell debris. The specific growth rate in the steady state was estimated

Table 2. Continuous (Chemostat) Culture of NB1 Hybridoma Cells - Effect of Cell Feedback on Fermenter Productivity

	Dilution rate H^{-1} (D)	Viable cells in culture x 10^{-6}/ml	Viable cells in culture x 10^{-6}/ml	IgM mg/litre ($\tilde{P}$)	Fermenter productivity mg IgM/litre working volume/day (D$\hat{P}$ x 24)
Chemostat without cell feedback	0.020	1.2	1.2	30	14
Chemostat with cell feedback	0.042	5.8	0.06	75	76

indirectly (by measurement of metabolic quotients and comparison with data from conventional chemostats) and estimated to be 0.019 h^{-1}. It will be seen that the cell population was greatly increased and the productivity of the vessel per unit time was 5.4 fold higher. We are now evaluating the feasibility of this approach at the pilot (30 litre) scale.

Antibody Recovery and Purification

The first step in recovering antibody from a culture supernatant is to separate the product from cells. This clarification step can be achieved by filtration or centrifugation. The next problem is to recover the antibody from a large volume of culture fluid (up to 1000 litres) containing the product in relatively dilute solution. Our approach here has been to use a rapid concentration step using tangential flow ultrafiltration equipment. By this technique one can achieve at least a ten-fold concentration with virtually quantitative recovery of antibody within an hour of harvesting the fermenter.

The antibody is now concentrated to between one and five grams per litre and can be purified by a variety of techniques. Depending on the properties of the particular antibody we use a combination of precipitation, ion exchange chromatography, gel filtration and affinity chromatography to achieve purities of greater than 95%.

Figure 7 shows the profile of proteins in the culture supernatant and the effect of purification by a precipitation step followed by an ion exchange chromatography step.

It is apparent that antibody protein is a minor proportion of the total protein in the culture fluid at harvest. The contaminating proteins derive from animal serum which is commonly used to supplement growth media.

Serum Free Culture Media

The use of serum free media would facilitate the purification of monoclonal antibodies. In addition we can identify other advantages.

Elimination of foetal bovine serum (the serum supplement most commonly used) would remove the most expensive component of the culture medium. It would also remove the risk of introducing adventitious microorganisms via serum. Finally a medium of defined and consistent composition offers greater potential for optimising the process.

A number of serum substitutes have been described (Chang et al., 1980; Cleveland et al., 1983; Fazekas de St. Groth, 1983; Kawamoto et al., 1983; Kovar & Franek, 1983; Murakami et al., 1983) and a number of serum free media are now available commercially.

We have now designed a serum free medium which we have used to grow a wide range of hybridomas in both static and agitated culture. Figure 8 shows growth and antibody synthesis for an IgM synthesising mouse hybridoma growing in a 5 litre airlift fermenter in serum free culture medium. Antibody yield and kinetics of production are similar to that seen in medium containing serum.

This serum-free medium contains some protein (1g l^{-1}). For comparison foetal calf serum at 10% v/v contributes about 5 g protein l^{-1} of culture medium. For an average antibody yield of 100 mg l^{-1} the antibody would represent 9% and 2% respectively of the total protein in serum free and

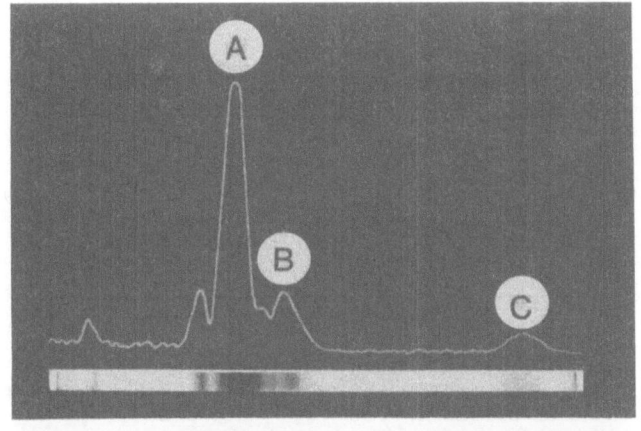

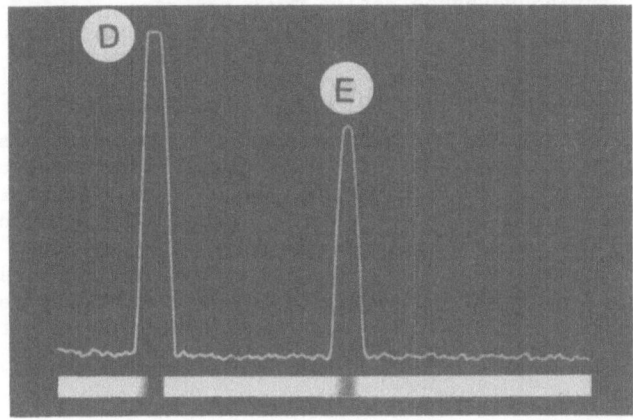

Fig. 7. Proteins contained in culture fluid and a
purified IgG preparation were separated by
SDS-PAGE (Laemmli, UK (1970), Nature 227,
680685) using a 10% (w/v) acrylamide gel.
Samples were reduced with 2-mercapto-
ethanol. After staining with coomassie
blue the separated proteins were scanned
with a microdensitometer and the scans
aligned with the gel tracks. The crude
culture fluid contained a majority of
protein other than IgG especially albumin
(Fig. 7, A). Immunoglobulin heavy and
light chain were easily identified (Fig. 7,
B and C respectively). Only heavy and
light chain bands (Fig. 7, D and E
respectively) were detected in the purified IgG preparation.

serum containing media. We have now developed a reduced protein medium
containing 10 mg 1^{-1} added protein, which has been used in small scale
culture and is now being evaluated at the pilot scale. In this medium, the
added proteins contribute less than 10% of the protein in the product.

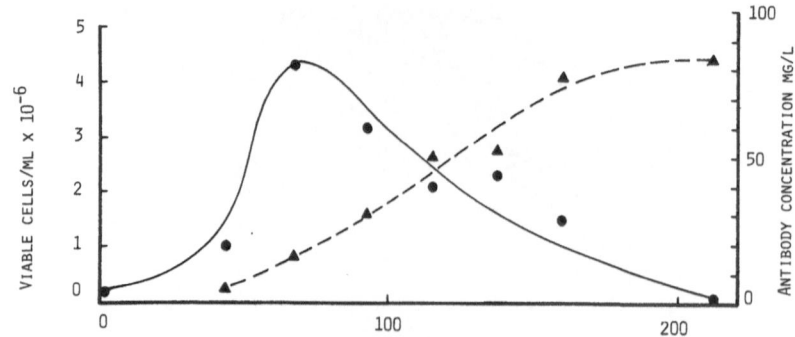

Fig. 8. Batch culture of a mouse hybridoma cell line, NB1.
Growing in serum-free medium in an airlift fermenter.
Viable cells/ml x 10^{-6} ————●————;
Antibody concentration mg/l ————▲————.

SUMMARY

We have developed a production process for monoclonal antibodies based
on the airlift reactor and capable of producing kilogram amounts. Efficient
recovery and purification processes have been devised which result in a
product which is typically better than 95% pure. The need to use expensive
animal serum in the culture medium is rapidly being overcome by the
development of serum free media. Looking to the future we can see great
scope for further optimisation of the cell culture process for making
antibodies, both by systematic study of cell physiology and by genetic
manipulation of the producing cell.

REFERENCES

Birch, J.R., Thompson, P.W., Lambert, K., Boraston, R., 1984, presented at
    the 188th American Chemical Society National Meeting, Academic Press,
    in press.
Boraston, R., Thompson, P.W., Garland, S. and Birch, J.R., 1983a, Develop.
    Biol. Stand., 55:103-111.
Boraston, R., Garland, S. and Birch, J.R., 1983b, J. Chem. Tech. Biotech.,
    33b: 200.
Chang, T.H., Steplewski, Y. and Koprowski, H., 1980, J. Immunol. Methods.,
    39:369-375.
Cleveland, W.L., Wood, I.and Erlanger, B.F., 1983, J. Immunol. Methods.,
    56:221-234.
Fazekas de St. Groth, S., 1983, J. Immunol. Methods., 57:121-136.
Fleischaker, R.J. and Sinskey, A.J., 1981, Eur. J. Appl. Microbiol.
    Biotechnol., 12:193-197.
Geyer, D.S., Collins, A.J., Koch, G.A. and Rupp, R.G., 1984, In vitro 20,
    No. 3, part 11, abstract 104.
Glacken, M.W., Fleischaker, R.J. and Sinskey, A.J., 1983, Trends in
    Biotechnology, 1:102-108.
Goding, J.W., 1980, J. Immunol. Methods, 39:285-308.
Hill, C.R., Birch, J.R. and Benton, C., 1984, in: "Bioactive Microbial
    Products III - Downstream Processing. Special publication of the
    Society for General Microbiology", Academic Press, London and New
    York.
Katinger, H.W.D., Scheirer, W. and Kromer, E., 1979, Ger. Chem. Eng., 1:
    31-38.
Kawamoto, T., Sato, J.D., Le, A., McClure, D.B. and Sato, G.H., 1983, Anal.
    Biochem., 130:445-453.

Kohler, G. and Milstein, C., 1975, Nature, 245:495.

Kovar, J. and Franek, F., 1984, Immunology Letters, 7:339-345.

Lydersen, B.K., Pugh, G.G., Paris, M.S., Sharma, B.P. and Noll, L.A., 1985, Bio/Technology, January issue, pp 63-67.

Mach, J-P., Buchegger, F., Forni, M., Ritschard, J., Berche, C., Lumbroso, J-D., Schreyer, M., Giradet, C., Accola, R. and Carrel, S., 1981, Immunology Today, December issue, pp 239-249.

Marcipar, A., Henno, P., Lentwojt, E., Roseto, A. and Brown, G., 1983, Ann. N.Y. Acad. Sci., 413: 416-420.

Marx, J.L., 1982, Science, 216:283-285.

Miller, R.A., Maloney, D.G., Warnke, R. and Levy, R., 1982, The New England Journal of Medicine, 306: 517-522.

Murakami, H., Edamoto, T., Shinohara, K. and Omura, H., 1983, Agric. Biiol. chem., 47:1835-1840.

Nilsson, K., Scheirer, W., Merten, O.W., Ostberg, L., Liehl, E., Katinger, H.W.D. and Mosbach, K., 1983, Nature, 302:629-630.

Nowinski, R.C., Tam, M.R., Goldstein, L.C., Stong, L., Kuo, C-C., Corey, L., Stamm, W.E., Handsfield, H.H., Knapp, J.S.and Holmes, K.K., 1983, Science, 219: 637-643.

Pirt, S.J., 1985, "Principles of Microbe and Cell Cultivation", Blackwell Scientific Publications.

Pullen, K., Johnston, M.D., Phillips, A.W., Ball, G.D. and Finter, N.B. 1984, Proceeding of the ESACT/IABS meeting, May 1984, Develop. Biol. Stand., in press.

Sacks, S.H. and Lennox, E.S., 1981, Vox Sanguinis, 40:99-104.

Secher, D.S. and Burke, D.C., 1980, Nature, 285:446-450.

Tovey, M.J., 1980, Advances in Cancer Research, 33:1-37.

Vitetta, E.S., Krolick, K.A., Miyana-Inaba, M., Cushley, W. and Uhr, J.W. 1983, Science, 219:644-650.

Voak, D. and Lennox, E.S., 1983, Biotest Bulletin, 4:281-290.

Wiemann, M.C., Ball, E.D., Fanger, M.W., Dexter, D.L., McIntyre, O.R., Bernier, Jr., G. and Calabresi, Pl., 1983, Clin. Res., 31:511.

EFFECT OF MOLECULAR HYDROGEN ON ACETATE DEGRADATION BY METHANOSARCINA

BARKERI 227

David R. Boone and Robert A. Mah

Division of Environmental and Occupational Health Sciences
School of Public Health, University of California
Los Angeles, CA 90024

SUMMARY

Acetate is the ultimate intermediate for over two-thirds of the methane produced by anaerobic digestors, and methanosarcinae are the predominant aceticlastic methanogens, especially in high-rate digestors. The remainder of the methane produced in anaerobic digestors (about one third) comes from the oxidation of hydrogen and the concomitant reduction of carbon dioxide to form methane. Hydrogen concentrations control many important catabolic reactions in anaerobic digestion, especially those involved in hydrogen production and degradation. Because hydrogen concentration is normally very low (several tenths of a micromole) and its turnover is very rapid, the entire pool size is degraded many times per second, and so minute changes in the production or degradation rate of hydrogen may be instantaneously translated into fluctuations in its concentration. Also, hydrogen concentration may not be uniform in the digestor. It is likely that there are microenvironments in anaerobic digestors having higher and lower concentrations than the average since hydrogen production is not uniform, and particles may account for a major portion of the hydrogen production. These fluctuations and the presence of microenvironments may be important in the inhibition of acetate degradation by Methanosarcina barkeri.

Acetate degradation by Methanosarcina barkeri (the type species), as well as other Methanosarcina species, is inhibited by elevated levels of molecular hydrogen. This inhibition appears to be a typical catabolite repression. Another kind of hydrogen inhibition has been noted also: pure cultures which were grown on hydrogen are not able to use acetate when inoculated into media containing acetate as the sole catabolic substrate.

We investigated catabolite repression of acetate degradation by growing Methanosarcina barkeri on acetate and providing various levels of hydrogen. Syntrophomonas wolfei produces hydrogen during growth on butyrate, but hydrogen can be produced only when its concentration is very low (near that found in digestors). M. barkeri 227 was unable to use hydrogen produced by this bacterium. We found that hydrogen at concentrations near those found in digestors neither inhibited nor stimulated acetate degradation by Methanosarcina barkeri 227. Under these conditions this organism did not appear to utilize hydrogen. At higher hydrogen concentrations M. barkeri 227 was able to use hydrogen, but as long as hydrogen concentration was elevated acetate degradation remained inhibited.

Another type of inhibition of acetate degradation was investigated. Hydrogen-grown cultures of <u>Methanosarcina barkeri</u> are unable to utilize acetate except after a prolonged lag-phase (several weeks) and sometimes not even then. We have found that, when <u>M. barkeri 227</u> cells were grown on a combination of hydrogen and acetate that hydrogen was utilized first. When hydrogen was exhausted, acetate degradation commenced with no inhibition. If acetate was absent and these cultures were used to inoculate media with acetate as substrate, a prolonged inhibition normally resulted. We also found that at the beginning of or after the onset of this inhibition, the addition of small amounts of hydrogen relieved the inhibition, and that methanol or trimethylamine addition was less efficient at relieving inhibition.

INTRODUCTION

Acetate accounts for two-thirds or more of the methane produced from anaerobic digestors (Boone, 1982; Smith and Mah, 1966). The production and degradation of acetate there proceed simultaneously, so that the concentration of acetate normally remains low. Thus the continuous degradation of acetate is crucial to normal functioning of anaerobic digestors for two major reasons: (a) because such a large fraction of the methane produced comes from acetate, high yields of methane depend on complete conversion of acetate; (b) in the absence of continuous acetate removal by its degradation, acetate would accumulate. Analysis of the buffer systems of anaerobic digestors indicates that when acetic acid is produced by bacterial fermentation, it ionizes to acetate, producing protons. This tends to lower with pH. As acetate is degraded to methane and carbon dioxide, protons are consumed, tending to raise the pH. Bicarbonate is the major buffering system of digestors, and since the bicarbonate concentration is normally about 0.05 M it can easily be overcome by acetate accumulation unless accumulation is balanced by degradation. Empirical studies have shown that when methanogenesis is inhibited, the pH of digestors usually drops. This drop in pH can permanently disable the fermentation, so that digestors may have to be emptied and restarted with fresh inoculum. Thus, in the control of anaerobic digestion, the aim should be to manipulate fermentation parameters to maximize the ability of methanogenic bacteria to degrade acetic acid, and to provide them (via microbial fermentation) with that substrate at a rate just below their maximum rate of degradation.

Essentially all of the methane produced from sources other than acetate comes from molecular hydrogen (Boone, 1982; Boone, 1984; Mackie and Bryant, 1981). Although hydrogen is not quantitatively as important as acetate in terms of methane produced, it may be more important in other ways (Boone, 1982; Boone, 1985a; Bryant, 1979). Hydrogen concentrations in healthy digestors are very low (less than 0.5 µM) and turn over very rapidly (Boone, 1985a). Small changes in hydrogen concentration may control the rates of many of the reactions of anaerobic digestion, including fermentation of amino acids and carbohydrates, degradation of volatile organic acids longer than acetate (e.g. propionate [6], butyrate [12], and long-chain fatty acids), and the production of volatile organic acids from hydrogen, carbon dioxide, and acetate (Bryant, 1979). It has also been established that acetate degradation in pure (Mah et al., 1978; Smith and Mah, 1978) and mixed cultures (Baresi et al., 1978) may be inhibited by elevated hydrogen concentrations. Methanosarcina are the dominant acetate-degrading methanogens present in high-rate anaerobic digestors; thus the present study was initiated to examine in detail the effects of various hydrogen concentrations on the acetate metabolism of an important strain of acetate-degrading methanogen, <u>Methanosarcina barkeri 227</u>.

RESULTS AND DISCUSSION

Effects of very low Hydrogen Concentrations on Methanosarcina barkeri 227

We wished to test the degradation of acetate by M. barkeri 227 in the presence of dissolved hydrogen concentrations of about 0.2 μM. This is equivalent to about 0.005% of the gas phase, and such levels would be very difficult to add and maintain by bubbling hydrogen-containing gas into the culture medium. We presently do not have the instrumentation to measure such quantities, and hydrogen goes into solution very slowly. Therefore we used the bacterium Syntrophomonas wolfei to produce hydrogen. This bacterium (McInerney et al., 1979) oxidizes butyrate to acetate, producing hydrogen. The thermodynamics of this reaction allow it to proceed only when hydrogen concentrations are low. There are also limits on the minimum hydrogen concentration which methanogens can utilize, so that there is a narrow range of hydrogen concentrations which allows both butyrate degradation and methanogenesis for hydrogen (Boone, 1985b). This range is near the normal hydrogen concentration found in anaerobic digestors, and so co-cultures of Syntrophomonas wolfei and hydrogen-utilizing methanogens were used to generate a stable, low hydrogen concentration.

A co-culture of Syntrophomonas wolfei and Methanospirillum hungatei was grown in medium with 20 mM butyrate as substrate. This culture was co-inoculated with Methanosarcina barkeri 227 and incubated at 37°C. Acetate concentration remained low during growth, and the culture was transferred (50% transfer) as soon as butyrate was depleted. When butyrate in this culture was depleted the culture was used as inoculum (10% transfer) for media containing 50 mM acetate. In some bottles 5 mM butyrate was also added, which would provide a source from which Syntrophomonas wolfei could produce hydrogen. We found that the presumed presence of hydrogen from this source did not affect the rate of acetate degradation by Methanosarcina barkeri and that in bottles containing butyrate that butyrate was degraded, indicating that hydrogen was produced (Figure 1). Microscopic examination of the inoculum showed large numbers of Methanospirillum hungatei, indicating that it successfully outcompeted Methanosarcina barkeri for the low levels of hydrogen. This was expected, as Methanosarcina barkeri has a higher K_m for hydrogen than do non-acetotrophic methanogens. We also tested the effect of even lower concentrations of hydrogen by using Desulfovibrio strain G11 to scavenge the hydrogen. We omitted butyrate from the media and included sulfate. Since Desulfovibrio strain G11 is able in the presence of sulfate to use hydrogen at extremely low levels (significantly lower than hydrogen-utilizing methanogens), the addition of sulfate should keep hydrogen levels even lower, perhaps even metabolizing small amounts of hydrogen produced by Methanosarcina. (During growth on acetate, Methanosarcina strain TM-1 and Methanosarcina acetivorans produces and maintains low concentrations (<0.1% of the gas phase) of hydrogen (Lovley and Ferry, 1985) and we have observed that Methanosarcina barkeri 227 produces small amounts of hydrogen, just at the levels of detection (ca. 0.1 to 0.01%)). We found (Figure 2) a slight stimulation of methanogenesis from acetate when sulfate was included, suggesting that hydrogen removal might be stimulatory. However, there are alternate hypotheses which also could explain these data. The growth of the Desulfovibrio may stimulate Methanosarcina by cross-feeding some stimulating nutrient, or Methanosarcina may be stimulated by sulfide produced by the Desulfovibrio. It is also possible that in the absence of sulfate Desulfovibrio strain G11 produced small amounts of hydrogen from the organic constituents of the medium (yeast extract and Trypticase peptone), and that the effect of sulfate was to eliminate this source of hydrogen. The possibility that hydrogen removal enhances growth of Methanosarcina barkeri 227 was supported by other data. The presence of Methanospirillum hungatei, which cannot metabolize Trypticase peptone or yeast extract, also stimulated methanogenesis. The

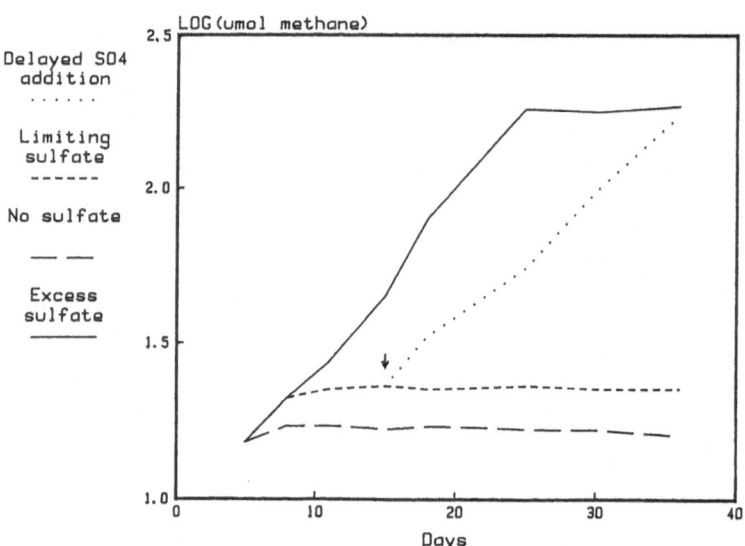

Fig. 1.   Ability of M. barkeri 227 to metabolize very
          low levels of H_2. Tricultures of M. barkeri,
          Syntrophomonas wolfei, and Desulfovibrio sp. G11
          were grown on media with butryate and various
          amounts of sulfate. When sulfate was limiting,
          methanogenesis from butryate was retarded. When
          sulfate was replenished (at the arrow excess was
          added) methanogenesis resumed.

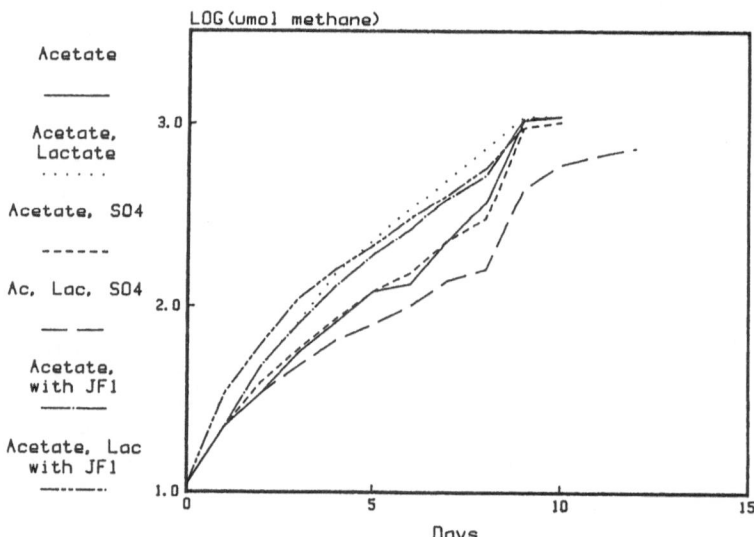

Fig.2. Growth of M. barkeri 227 on acetate. Co-cultures
        were grown with Desulfovibrio G11; lactate (from
        which strain G11 can produce H_2) or sulfate (with
        which strain G11 can utilize H_2) were added. Where
        indicated an alternate H_2-oxidizing methanogen,
        Methanospirillum hungatei JF1 was also added.

differences in the rates were small, but they may indicate that acetate degradation by <u>Methanosarcina barkeri</u> is stimulated by hydrogen removal from growing cells.

The above experiment showed that low levels of hydrogen did not inhibit acetate degradation by <u>Methanosarcina barkeri</u>, but they did not show whether hydrogen and acetate could be utilized simultaneously. This was tested by using a co-culture of <u>Syntrophomonas wolfei</u> and <u>Desulfovibrio</u> strain G11. (<u>Desulfovibrio</u> strain G11 uses hydrogen to reduce sulfate to sulfide; hydrogen utilization depends on the presence of sulfate.) This co-culture was grown in medium with 20mM butyrate and 20 mM sulfate, and then co-inoculated with an acetate-grown culture of <u>Methanosarcina barkeri</u> (10% inoculum of each) into media with 20 mM butyrate and various amounts of sulfate. In the absence of sulfate, <u>Desulfovibrio</u> strain G11 would be unable to utilize the hydrogen produced from butyrate by <u>Syntrophomonas wolfei</u>. Thus hydrogen would accumulate and inhibit acetate production from butyrate by <u>Syntrophomonas wolfei</u> unless these low levels (ca. 10^{-5} to 10^{-4} atmospheres) could be used by <u>Methanosarcina barkeri</u>. Figure 1 shows that when excess amounts of sulfate were added, methane production from acetate was complete. When no sulfate was added, very little methane was produced. When sulfate concentration was limiting, methanogenesis was limited. Thus <u>Methanosarcina barkeri</u> was not able to utilize hydrogen at the concentration produced by <u>Syntrophomonas wolfei</u>, even without competition. When, later, small amounts of hydrogen, sulfate, or acetate were added to bottles which initially had limiting amounts of sulfate (Figure 1), methanogenesis resumed. Presumably these additions had the following effects: (a) Hydrogen addition increased hydrogen concentration to a level which <u>Methanosarcina barkeri</u> was able to use. When the hydrogen was gone, methanogenesis ceased. Thus <u>Methanosarcina barkeri</u> was able to use hydrogen, but not at very low levels. (b) Sulfate addition allowed <u>Desulfovibrio</u> strain G11 to use hydrogen, thus allowing <u>Syntrophomonas wolfei</u> to convert more butyrate to acetate. The acetate served as substrate for <u>Methanosarcina barkeri</u>. (c) Acetate added was directly used by <u>Methanosarcina barkeri</u>. This indicated that <u>Methanosarcina barkeri</u> could not catabolize hydrogen at the concentrations normal to healthy digestors. However, it is possible that hydrogen may still be used for assimilatory purposes under these conditions. It is also possible that higher hydrogen concentrations may be used simultaneously with acetate as catabolic substrates. These higher concentrations may occur in digestors either during transition periods (batch fed digestors or during upsets) or in micro-environments with elevated hydrogen concentrations (Boone, 1984; Boone, 1985b).

Acetate Use by Hydrogen-Grown Inocula

When <u>Methanosarcina barkeri 227</u> is grown on media containing hydrogen and carbon dioxide as the energy substrate, it does not utilize acetate when transferred to media with acetate as the only energy substrate (Mah et al., 1978). When we repeated this experiment, we found that in many cases acetate degradation started after only a short lag (2 to 5 days). To obtain the inoculum for this experiment we took acetate-grown cultures, and transferred them to media with hydrogen as the only catabolic substrate. After growth was nearly complete this culture was used as inoculum for the study, in which media containing acetate and various amounts of hydrogen were inoculated.

The high rate at which cultures in acetate-only media grew (Table 1) appeared to be in contradiction to earlier work from this laboratory. This led us to examine more closely the inoculum used for the experiments. The earlier study used an inoculum maintained for many transfers on medium with hydrogen only. Since then we have been maintaining stock cultures of

Table 1. Inhibition of Acetate Degradation in Cultures of <u>Methanosarcina barkeri 227</u> by Pre-growth in Media Containing Hydrogen and Carbon Dioxide as the Energy Substrate[a]

Substrates		Number[b]	Methane produced (μmol/vessel)	Days until maximum methane
Acetate (mM)	Others[b] (mM^c)			
50	none	3/5	194	28
		1/5	54	28[d]
		1/5	3	23
50	methanol (1)	4/5	208	21
		1/5	32	19[e]
50	TMA (1)	4/5	208	25
		1/5	112	28[d]
50	hydrogen (4.1)	4/5	78	28[d]
		1/5	226	28
50	hydrogen (8.2)	3/4	104	28[d]
		1/4	223	24
50	hydrogen (40.9)	3/5	220	26
		2/5	91	28[d]
50	hydrogen (81)	3/4	258	27
		1/4	/39	28[d]

[a]Acetate-grown <u>Methanosarcina barkeri 227</u> cultures were grown once in hydrogen media and used to inoculate tubes with acetate as the only substrate.

[b]For each treatment, replicate vessels were grouped according to their response (e.g. all vessels in which acetate degradation occurred were averaged); the numbers shown are the number in the group over the total replicates for each treatment.

[c]Alternate substrates were added after 14 days incubation with acetate as the sole substrate. Amounts for hydrogen are millimoles of hydrogen gas per liter of culture medium.

[d]The experiment was terminated after 28 days, and these vessels were still producing methane at that time.

[e]These vessels were producing methane but the medium became oxidized, killing the cultures.

<u>Methanosarcina barkeri 227</u> only on acetate, since we found that these cultures could grow on the other substrates (hydrogen and carbon dioxide, or methanol or methylamines) without a lag. Therefore we took the acetate-grown culture, and inoculated hydrogen-only medium. When growth was nearly complete (based on gas analysis) the culture was transferred (5% inoculum) again to hydrogen-only medium. This was repeated six times, and after each transfer four cultures were inoculated with acetate as sole substrate. The experiment is not yet complete, but preliminary data suggest that under

these conditions a minium of three transfers on hydrogen-only media were necessary in order to demonstrate inhibition of acetate degradation. The fourth transfer was used for inoculum to repeat the above experiment demonstrating relief of acetate-inhibition by the addition of hydrogen. This time we found acetate utilization in acetate-only bottles to be rare (Table 2). Hydrogen appeared to be the most effective substrate for relieving the inhibition of acetate degradation; as little as 1 ml of hydrogen gas added per bottle (20 ml liquid, 44 ml gas headspace) was sufficient to relieve the inhibition. This was the smallest amount of hydrogen tested. Figure 3 shows the time course of methane production from these tubes. We are currently in the process of repeating these experiments, but we are also testing smaller amounts of added hydrogen (0.5, 0.1, 0.05 ml per bottle) and also trimethylamine as alternate substrate.

Table 2. Inhibition of Acetate Degradation in Cultures of <u>Methanosarcina barkeri 227</u> by Pre-growth for Four Transfers in Media Containing Hydrogen and Carbon Dioxide as the Energy Substrate[a]

Substrates		Number[b]	Methane produced (μmol/vessel)	Days until maximum methane
Acetate (mM)	Others[b] (mM[c])			
50	none	6/7	1	5
		1/7	482	35[d]
50	methanol (1)	3/4	1031	24
		1/4	136	35[d]
50	methanol (5)	3/4	666	35[d]
		1/4	1045	18
50	hydrogen (2)	4/4	979	17
50	hydrogen (10)	4/4	1010	17
50	hydrogen (20)	4/4	1016	13

[a]Acetate-grown <u>Methanosarcina barkeri 227</u> cultures were grown for transfers in hydrogen media and used to inoculate tubes with acetate as the only substrate; culture vessels contained 20 ml medium, compared to 5 ml in Table 1.

[b]For each treatment, replicate vessels were grouped according to their response (e.g. all vessels in which acetate degradation occurred were averaged); the numbers shown are the number in the group over the total replicates for each treatment.

[c]Alternate substrates were added before inoculation (for Table 1 they were added after 14 days incubation); amounts for hydrogen are millimoles of hydrogen gas per liter of culture medium.

[d]The experiment was terminated after 35 days, and these vessels were still producing methane at that time.

Fig. 3. (A) Acetate-grown <u>Methanosarcina barkeri 227</u> cultures were grown for four transfers (2.5% inoculum) in media with H_2-CO_2 as sole catabolic substrate and inoculated into media with acetate as sole catabolic substrate. One culture out of seven used the acetate. (B) Effect of methanol on the ability of hydrogen-grown <u>Methanosarcina barkeri 227</u> to use acetate. Culture conditions were the same as Figure 3(A) except methanol was added to some vessels. Cultures with added methanol slowly reacquired the ability to use acetate, compared with controls without added methanol. (C) Effect of hydrogen on the ability of hydrogen-grown <u>Methanosarcina barkeri 227</u> to use acetate. Culture conditions were the same as Figure 3(A) except hydrogen was added to some vessels. Cultures with added hydrogen quickly reacquired the ability to utilize acetate.

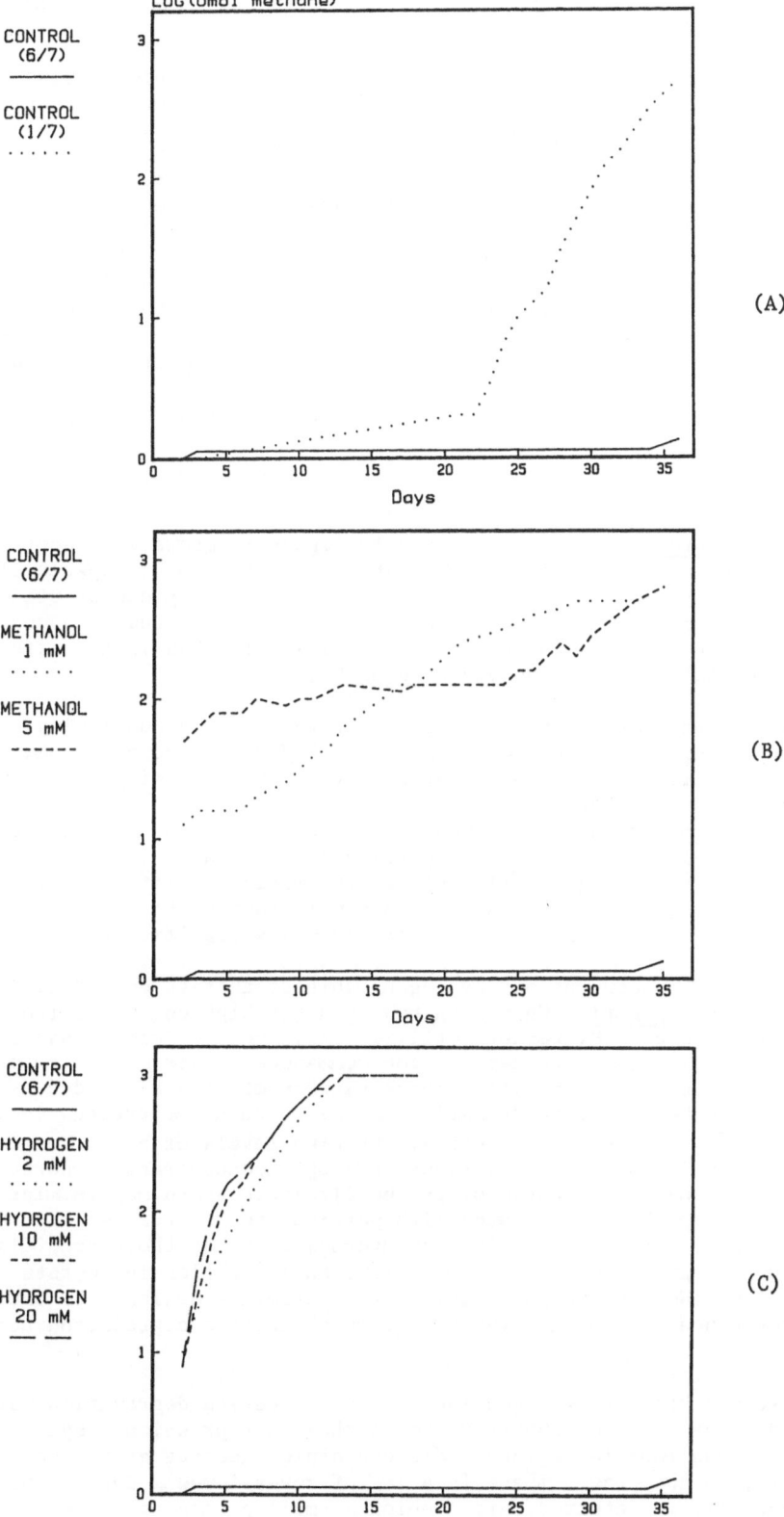

Determination of Minimum Hydrogen Concentration which Causes Inhibition of Acetate Degradation

We have shown that low hydrogen concentrations (concentrations common to healthy digestors) did not inhibit acetate degradation and that high concentrations did. We repeated studies done in other laboratories (McInerney and Bryant, 1981; Traore et al., 1983) testing interspecies hydrogen-transfer with lactate as substrate. The two species we used were Desulfovibrio strain G11, which used lactate and produced acetate, carbon dioxide and hydrogen, and Methanosarcina barkeri 227, which used hydrogen to reduce carbon dioxide and produce methane. We found that these two species did grow together in batch culture, and that acetate accumulated (apparently until lactate was exhausted, based on the amount of methane produced from hydrogen); then acetate was degraded to methane. Thus the level of hydrogen produced during rapid growth of Desulfovibrio strain G11 was too high to allow acetate degradation by Methanosarcina. We intend to pursue this by performing continuous culture studies with these two species.

CONCLUSIONS

Methanosarcina barkeri in anaerobic digestors grows with acetate as the major methanogenic substrate. Although this bacterium is capable of catabolizing hydrogen to produce methane, it apparently cannot utilize the low concentrations present in digestors which are operated on a continuous basis. Molecular hydrogen at these low levels also has little effect on acetate degradation by Methanosarcina barkeri.

When Methanosarcina barkeri 227 is grown on media containing acetate as the only energy substrate, small amounts of hydrogen are produced. The reasons for this are not known, but we have tested the effect of removing this hydrogen and maintaining lowered hydrogen concentrations by co-culturing M. barkeri with hydrogen-utilizing methanogens or hydrogen-utilizing desulfovibrio. These experiments showed a very small increase in growth rate with hydrogen utilization. The measured difference in growth rate was very small, and the experiments will have to be refined and repeated to determine whether the difference was significant.

At higher concentrations hydrogen inhibits acetate degradation by Methanosarcina barkeri. When both hydrogen (at high concentration) and acetate are present, M. barkeri utilizes only hydrogen until that substrate is depleted. Then acetate degradation commences. This may be important in batch fed digestors, where after large batches of feed are added hydrogen concentration may increase dramatically. Also, when methanogenesis is inhibited in a digestor, for instance by high levels of heavy metals or addition of chlorinated hydrocarbons, hydrogen concentration increases. Also, even in healthy, continuously fed digestors there may be micro-environments, perhaps associated with particulate substrates, that have higher than normal concentrations of hydrogen. Under these conditions, elevated hydrogen concentration may cause an inhibition in acetate degradation. We have begun a series of experiments designed to determine the exact concentration of hydrogen at which acetate degradation becomes inhibited.

A second kind of hydrogen-inhibition of acetate degradation was observed which does not appear to be catabolite repression. Hydrogen-grown cultures, when inoculated into media containing acetate as the sole energy substrate, do not grow. There is a lag of several weeks during which no methanogenesis is detected, after which a small percentage of cultures began growth on acetate. If hydrogen is added to the acetate media, growth occurs, or if hydrogen is added after the inhibition is noted, the inhibition is relieved. The smallest amounts of hydrogen we tested (2.2% of the gas phase) are sufficient for reversal of inhibition, and we are currently testing smaller amounts.

REFERENCES

Baresi, L., R.A. Mah, D.M. Ward and Kaplan, I.R., 1978, Methanogenesis from acetate: enrichment studies, Appl. Environ. Microbiol., 36:186-197.
Boone, D.R., 1982, Terminal reactions in the anaerobic digestion of animal waste. Appl. Environ. Microbiol., 43:57-64.
Boone, D.R., 1984, Mixed-culture fermentor for simulating methanogenic digestors, Appl. Environ. Microbiol., 48:122-126.
Boone, D.R., 1985a, Fermentation reactions of anaerobic digestion, in: "Biotechnology Handbook", P.N. Cheremisinoff and R.P. Ouellette, eds., Butterworth/Ann Arbor Science Pub. (in press).
Boone, D.R., 1985b, Thermodynamics of catabolic reactions in the anaerobic digestor, in: "First Symposium on Biotechnological Advances in Processing Municipal Wastes for Fuels and Chemicals", A.A. Antonopoulos, ed., Academic Press (in press).
Boone, D.R. and Bryant, M.P., 1980, Proprionate-degrading bacterium, Syntrophobacter wolinii sp.nov., from methanogenic ecosystems, Appl. Environ. Microbiol., 40:626-632.
Bryant, M.P., 1979, Microbial methane production - theoretical aspects, J. Anim. Sci., 48:193-201.
Lovley, D.R. and Ferry, J.G., 1985, Production and consumption of H_2 during growth of Methanosarcina spp. on acetate, Appl. Environ. Microbiol., 49:247-249.
Mackie, R.I. and Bryant, M.P., 1981, Metabolic activity of fatty acid-oxidizing bacteria and the contribution of acetate, propionate, butyrate and CO_2 to methanogenesis in cattle waste at 40 and 60°C, Appl. Environ. Microbiol., 41:1363-1373.
Mah, R.A., Smith, M.R. and Baresi, L., 1978, Studies on an acetate-fermenting strain of Methanosarcina, Appl. Environ. Microbiol., 35:1174-1184.
McInerney, M.J., and Bryant, M.P., 1981, Anaerobic degradation of lactate by syntrophic associations of Methanosarcina barkeri and Desulfovibrio species and effect of H_2 on acetate degradation. Appl. Environ. Microbiol., 41:346-354.
McInerney, M.J., Bryant, M.P. and Pfennig, N., 1979, Anaerobic bacterium that degrades fatty acids in syntrophic association with methanogens. Arch. Microbiol., 122:129-135.
Smith, P.H. and Mah, R.A., 1966, Kinetics of acetate metabolism during sludge digestion, Appl. Microbiol., 14:368-371.
Smith, M.R. and Mah, R.A., 1978, Growth and methanogenesis by Methanosarcina strain 227 on acetate and methanol, Appl. Environ. Microbiol., 36:870-879.
Traore, A.S., Fardeau, M.-L., Hatchikian, C.E., LeGall, J. and Belaich, J.-P., 1983, Energetics of growth of a defined mixed culture of Desulfovibrio vulgaris and Methanosarcina barkeri: interspecies hydrogen transfer in batch and continuous cultures. Appl. Environ. Microbiol., 46:1152-1156.

IN VITRO INDUCTION OF NITROGENASE ACTIVITY IN FREE-LIVING RHIZOBIA BY A

NON-NODULATING LEGUME

M.A. Martins-Loução[1] and C. Rodriguez-Barrueco

Unit of Nitrogen Fixation, Centro de Edafologia y Biologia
Aplicada, CSIC, Salamanca, Spain
[1]Department of Plant Biology, Faculty of Sciences
1294 Lisboa Codex, Portugal

ABSTRACT

The effect of cultured carob cell extracts and exudates on acetylene reduction of three different species of Rhizobium, namely R. japonicum IS-33, "cowpea" Rhizobium TAL 169 and R. leguminosarum TAL 619, was studied. The exudates, excreted from plant callus cultures on an agar defined medium promote a synergistic activation of nitrogenase in pure cultures of rhizobia. "Cowpea" Rhizobium nitrogenase activity was stimulated by a high concentration of sucrose in the medium. Ability of exudates to influence Nase of R. leguminosarum and R. japonicum, respectively, increased in the presence of succinate. Stimulation could also take place when bacteria were grown on medium to which a crude cell extract was added. When the extract was incubated with rhizobia in a minimal nutrient medium a strong synergistic activation of R. leguminosarum and R. japonicum nitrogenase was observed. On this medium the cell extract showed an inhibitory effect on "cowpea" Rhizobium nitrogenase activity.

INTRODUCTION

All crops require large quantities of nitrogen for growth. This element which constitutes 80% of the atmosphere is very often considered to be the major soil nutrient limiting the primary productivity of several ecosystems. Enrichment of soil nitrogen in natural ecosystems is largely attributed to the biological nitrogen fixation process, carried out by microorganisms both under free-living conditions and in symbiosis with higher plants.

The knowledge of the fact that legumes in association with rhizobia can utilize atmospheric nitrogen is relatively recent as compared to the long period during which these plants have been cultivated for food production and soil fertility enrichment.

The formation of nodules on the nitrogen-fixing leguminous host-plants was considered a characteristic of a purely taxonomic importance. An erroneous opinion associating legumes and nodule formation is still known to exist, in spite of the fact that all members of the family Leguminoseae are not capable of forming nodules. Most of the members of the sub-family Caesalpinoideae are not known to possess nodules.

Such is the case of carob (<u>Ceratonia siliqua</u> L.), a tree reaching over 10 m. in height and having rich, glossy, evergreen foliage. The pods (10-30 cm long) contain 5-15 very hard seeds embedded in a sweet, mealy pulp. The tree is less exacting in its soil requirements than most fruit trees. Although carob can grow well in poor soils, no one has demonstrated as yet that it nodulates and is able to obtain nitrogenous nutrients by nitrogen fixation (Allen and Allen, 1981). Carob has a great importance in the economy of mediterranean countries due to the value of its products which constitute the basic raw material of several important industries such as Tate and Lyle, U.K. It was also considered as an important crop amongst the tropical and sub-tropical legumes by the National Academy of Sciences (NAS, 1979). Owing to this we thought that concentration ought to be given to this species in order to understand the causes governing the inability of this crop to form an effective nitrogen-fixing symbiotic association.

The economic importance of the <u>Rhizobium</u>-legume symbiosis has over the years been reflected in the amount of effort devoted to the study of these bacterial-higher plant relationships. Unfortunately, many investigations of these systems have been limited by the complexity of the nodule and could not provide insights to elucidate the cause or causes underlying the very different nodulating abilities of members of the Leguminoseae. So, unravelling the subtleties and the mechanisms of symbiosis continues to be a research challenge. Attempts to simplify such relationships have led to the development, during the last twenty years, of <u>in vitro</u> techniques to study this association.

Several reports have demonstrated nitrogenase activity (C_2H_2) in certain strains of <u>Rhizobium</u> grown in pure culture (Keister, 1975; Pagan et al., 1975). This activity can be stimulated by cultured leguminous and non-leguminous plant cells (Scowcroft and Gibson, 1975; Werner, 1976; Götz, 1980), root extracts (Kurz and LaRue, 1975) and seeds (Anderson and Phillips, 1976). As it was demonstrated (Storey et al., 1979; DeMoranville et al., 1981), soybean cells in suspension cultures produce diffusible factors which induce or stimulate nitrogenase activity in several rhizobial species. An inhibition of nitrogenase activity in free-living rhizobia was found when grown in contact with snake bean callus (McComb et al., 1975).

The establishment of callus from carob explants in a synthetic medium has been achieved as a first step towards the understanding of the physiology of this plant tree species (Martins-Loução and Rodrigues-Barrueco, 1981). Nitrogenase activity of <u>R. leguminosarum</u> associated with carob callus cultures was obtained despite the endogenous ethylene production developed by callus (Martins-Loução and Rodrigues-Barrueco, 1982, 1983). In the present work we developed a technique where only exudates and cell extracts of callus cultures in the media were used to induce nitrogen fixation in free-living rhizobia as measured by acetylene reduction.

MATERIAL AND METHODS

Callus Cultures

All experiments were carried out using callus cells from excised carob hypocotyls (Martins-Loução and Rodriguez-Barrueco, 1981). Calluses were grown under weak fluorescent light giving 12 $\mu E.sec^{-1}.m^{-2}$ of energy, provided by Philips Daylight 45 W lamps in a 12h photoperiod. The growth temperature was 26±2°C. (Figure 1).

Culture of Rhizobia

Experiments were done with three different species of <u>Rhizobium</u>: <u>R. japonicum</u> IS-33, "cowpea" <u>Rhizobium</u> TAL 169 and <u>R. leguminosarum</u> TAL 619. Pure cultures were grown and maintained on yeast-extract mannitol agar medium (Vincent, 1970). Subculturing onto fresh agar medium was done after 15 to 20 days incubation.

Callus Exudates

Cultured carob callus was transferred onto 55 ml tubes containing 15 ml of GLH 1 - with 87 mM sucrose - or GLH 3 - with 44 mM sucrose plus 25 mM succinate. GLH media, which lacked mineral nitrogen, contained the following substances in mg/1: KH_2PO_4, 299.3; $CaCl_2.2H_2O$, 77,7; KCl, 70.1; $MgSO_4.7H_2O$, 73.9; $MnSO_4.H_2O$, 10; H_3BO_3, 5; $ZnSO_4.7H_2O$, 1; KI, 1; $CuSO_4.5H_2O$, 0.2; $Na_2MoO_4.2H_2O$, 1; $CoCl_2.6H_2O$, 1; inositol, 108; thiamine.HCl, 5; nicotinic acid, 5; pyridoxine.HCl, 5; $FeSo_4.7H_2O$, 15; Na_2EDTA, 20; glutamine, 730.8; NAA, 1; BAP, 2; agar, 12000. The pH was adjusted to 6.0 for these media (Martins-Loução and Rodriguez-Barrueco, 1982). Callus cells were grown for 15 days on these two media prior to extraction of exudates and prior to inoculation with bacteria on the same media. All operations were carried out under sterile conditions.

Callus Cell Extracts

After 15 days on GLH 3, callus was homogenized in a cold mortar with liquid GLH 3 without hormones and then centrifuged at 10,000xg for 20 min. to remove cell-wall debris. The supernatant fraction constituted the callus cell extract. Approximately 1 g of material was homongenized with 25 ml of liquid medium which produced a supernatant of 15 ml. The extract was then filter-sterilized through 0.45 μm filters before addition to induction medium.

Two types of media were used in induction of nitrogenase activity in rhizobia: the above-mentioned GLH 3 and SCH with the same composition of the induction medium used by DeMoranville and coworkers (1981) but

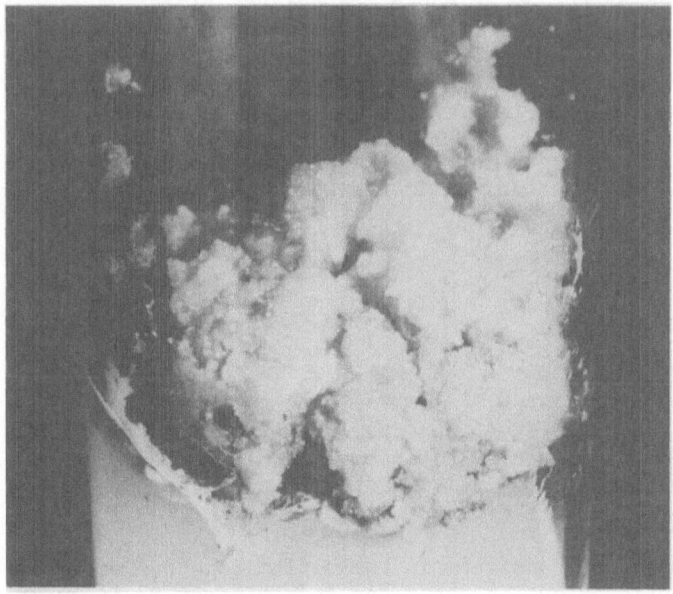

Fig 1. Callus hypocotyl from carob explants

containing the hormones NAA (1 mg/1) plus BAP (2mg/1). SCH contained the
following substances in mg/1: KH_2PO_4, 234; $CACl_2.2H_2O$, 75.5; KCl, 67.8;
$MgSO_4.7H_2O$, 34.5; $MnSO_4.H_2O$, 4.6; H_3BO_3, 1.5; $ZnSO_4.7H_2O$, 1.5;
$CuSO_4.5H_2O$, 0.04; $Na_2MoO_4.2H_2O$, 0.25; $CoCl_2.6H_2O$, 0.025; inositol, 120.7;
thiamine.HCl, 5; nicotinic acid, 5; pyridoxine.HCl, 5; glutamine, 292.3;
$FeSO_4.7H_2O$, 6.95; Na_2EDTA, 9.3; agar, 12000. The latter medium, with the
same substances as GLH but in lesser amounts, contained arabinose and
succinate as carbon sources.

Assay Procedures

The exudate experiments were carried out using 55-ml tubes containing
15 ml of GLH 1 or GLH 3 medium, respectively. The media which contained the
exudates from 15-day-old callus were inoculated with a loopful of a 2-week-
old culture of the different rhizobial species, and incubated in the dark at
28°C. Controls were rhizobia growing on pure GLH 1 and GLH 3 media.

The crude extract experiments were carried out using 100-ml Erlenmeyer
flasks containing 50 ml of either GLH 3 or SCH media. Four ml of filter-
sterilized extract were added to each of the flasks containing 50 ml of agar
medium at a temperature of 45°C. Controls were rhizobia growing on GLH 3 or
SCH media with 4 ml of filter-sterilized water.

For these experiments bacterial inocula were prepared in the following
manner: a 2-week-old agar slant culture was resuspended in 6 ml sterile
water and homogenized in a shaker. The resulting suspension of the
different species of Rhizobium contained approximately $4-6 \times 10^8$
bacteria/ml. Each culture flask was inoculated with 0.5 ml of this
bacterial suspension and incubated in the dark at 28°C.

After a 10-day-incubation period, the cotton plug was replaced by a
serum stopper and acetylene was injected to a concentration of 10%. After
24 h incubation, 1-ml gas samples of the enclosed atmospheres were withdrawn
with a 2-ml syringe and injected into a Varian Aerograph 2700 gas chromato-
graph equipped with a 1.5 m long alumina column, using standard techniques
(Hardy et al., 1968).

At the end of the experiments the culture flasks were opened and the
bacteria washed with 10 ml deionized water and centrifuged at 10,000 g for
15 min. Dry weights of the resulting pellet were determined by suspending
in 3 ml water and setting the sample to dry at 90°C for 24 h in preweighed
aluminium foil "boats".

Results of all experiments were calculated as nmols ethylene (g dry
weight)$^{-1}$. (24 h)$^{-1}$. Four replicates were included for each of the assays
here described.

RESULTS

Callus Exudates

The data show that the capacity of the rhizobia to develop nitrogenase
activity varies from species to species and it seems to be dependent largely
on the composition of the medium rather than on the presence of callus
exudates.

The presence of succinate in the medium together with carob exudates
had a synergistic effect on acetylene reduction by R. leguminosarum and
R. japonicum. On the contrary, carob exudates seemed to repress the

activity of "cowpea" Rhizobium on GLH 3. The nitrogenase activity of this species could be stimulated by callus exudates in the medium with sucrose as the only carbon source (Figure 2).

Callus Extracts

The carob cell extract incorporated into GLH 3 medium only presents a slight stimulation of rhizobial nitrogenase activity.

In order to determine whether the weak response of these rhizobia was due to an inhibitory factor of the extract or to an excess of nutrients, another medium was assayed. SCH had a smaller quantity of nutrients, containing arabinose as carbon source and half strength succinate.

The data presented in Figure 3 show that the addition of cell extracts to this medium much enhanced the activity shown by R. japonicum and R. leguminosarum. The activity of "cowpea" in pure culture was highly stimulated on this medium but the presence of callus extracts seemed to reduce such activity. These results appear to indicate that the stimulatory requirements for "cowpea" Rhizobium are not the same as for R. japonicum and R. leguminosarum.

DISCUSSION

Nitrogenase activity (acetylene reduction) by free-living R. leguminosarum, R. japonicum and "cowpea" Rhizobium, under the experimental conditions of this study, can be induced in the presence of callus exudates and cell extracts, respectively, although such activity also seemed to be largely dependent on the composition of the medium.

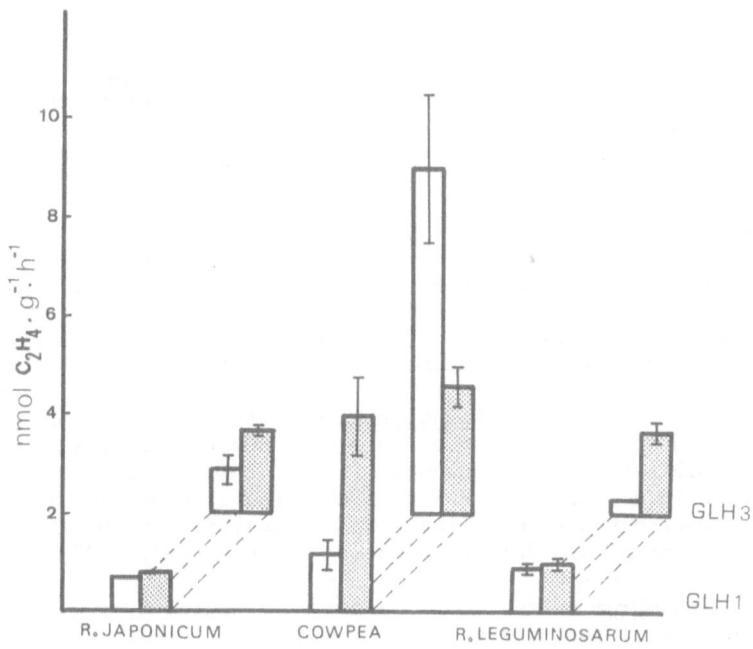

Fig.2 Nitrogenase activity by free-living Rhizobium in the presence of carob callus exudates. Each point represents the mean± SE from 4-5 replicates. White bars - control; dotted bars - Rhizobium + exudates.

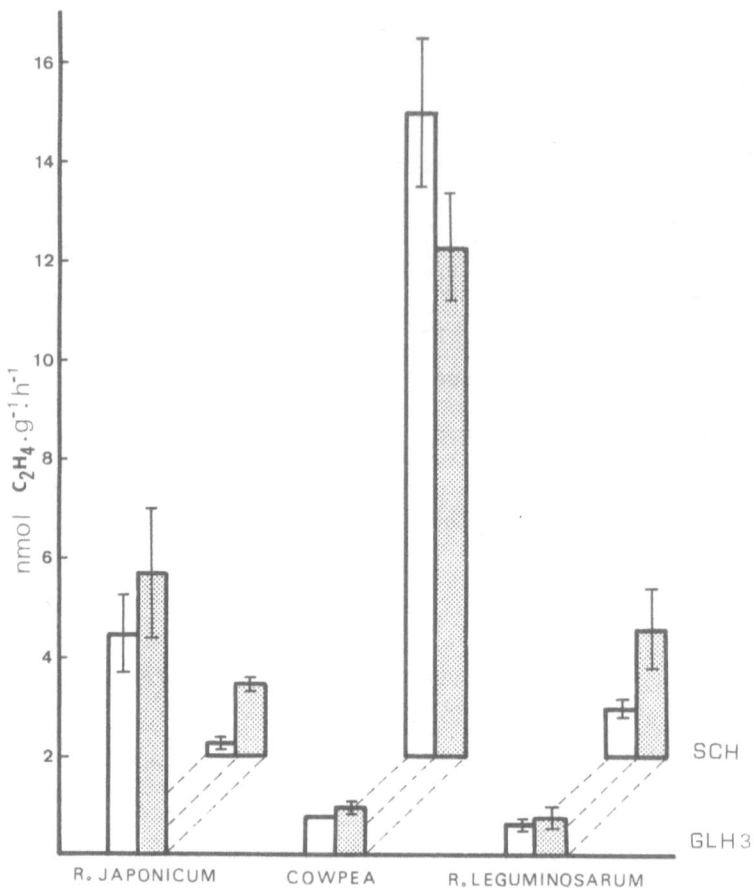

Fig. 3. Nitrogenase activity of free-living Rhizobium in the
presence of carob cell extracts. Each point represents the mean ± SE from 4-5 replicates. White
bars - control; dotted bars - Rhizobium + extracts

The different methodology employed in these experiments, in exudates
and cell extracts, do not allow us to compare the different degrees of
acetylene-reducing activity shown by rhibozia. The activity in the presence
of extracts was generally higher than in the presence of exudates, possibly
due to a higher concentration of nutrients or stimulatory substances in the
extracts, thus providing an enhancement of the activity.

The response of "cowpea" Rhizobium to exudates and extracts seems to
suggest that plant cells can also produce some substances that could
function as inhibitors. The occurence of plant cell inhibitors has also
been discovered in soybean cell extracts (DeMoranville et al., 1981) though
Reporter and coworkers have only found stimulatory factors of rhizobial
nitrogenase (Bednarski and Reporter, 1978; Storey et al., 1979).

"Cowpea" Rhizobium seemed to be much more affected by the composition
of the medium than by the presence of plant cells or exudates. However, an
increased acetylene-reducing activity of "cowpea" was obvious in the
presence of exudates when sucrose was the only carbon source (Figure 2).
Glenn and coworkers (1984) observed that R. leguminosarum maintains the
capacity to utilize certain sugars without apparently affecting the

induction of nitrogenase. This finding seems to be in agreement with that found in the present study for "cowpea" on a sucrose medium.

Succinate promotion of nitrogenase activity in Rhizobium pure cultures (McComb et al, 1975), soybean cell-Rhizobium associations (Child and LaRue, 1974) and bacteriod preparations (Bergersen and Turner, 1967), respectively, has been reported previously. In the present experiments succinate also promotes an enhancement of the rhizobial nitrogenase on "cowpea" pure cultures and on R. japonicum and R. leguminosarum in the presence of callus exudates. It has been suggested (Bergersen, 1977) that a relatively high concentration of succinate in the agar media is a requisite to create a low O_2 concentration necessary to permit the function of nitrogenase.

Glutamine concentrations higher than 5 mmoles inhibit growth of R. japonicum (Wilcockson and Werner, 1978) and the removal of glutamine can increase nitrogenase activity in Rhizobium strain 32H1 (Bergersen, 1977). The results outlined here with carob callus extracts agree with those findings. On SCH medium the concentration of glutamine is half that contained in GLH 3, providing an enhanced nitrogenase activity in R. japonicum and R. leguminosarum in the presence of callus extracts.

These experiments with free-living rhizobia and others with seedlings, excised roots and callus (Martins-Loução and Rodriguez-Barrueco, 1982; Martins-Loução, 1985), emphasize that the bacterial cell possesses the complete genome for the expression of nitrogenase and that plant cell substances have an inducing effect on that activity. At present, we do not know why Ceratonia siliqua, though a legume, is not a nodule-bearing plant. The demonstration that Rhizobium has a clearly independent nitrogen-fixation capability will intensify investigations on the relationships between the bacteria and the plant, specifically those involved in the infection process.

Indeed, the genetic constitution of the host which determines its susceptibility to infection by Rhizobium, may constitute yet another important factor for consideration. Experiments reported herein reveal the ability of carob cell extracts and exudates to induce nitrogenase activity in Rhizobium. This leads us to infer that the barriers posed by the plant to the establishment of a symbiotic association, intervene either at the stage of infection or nodule formation, rather than acting on the expression of the nitrogenase. Therefore, in comparing carob and other non-nodulating legumes with Parasponia, a member of Ulmaceae known to associate symbiotically with Rhizobium (Akkermans et al., 1978), one is confronted with the need to identify those specific genes controlling the establishment of a symbiotic association in general and specifially those characterizing the legume symbiosis, with the ultimate objective of enlisting the necessary prerequisites for the establishment of an effective nitrogen-fixing system.

REFERENCES

Akkermans, A.D.L., Abdulkadir, S. and Trinick, M.J., 1978. N_2-fixing root nodules in Ulmaceae: Parasponia or (and) Trema spp. ? Pl. Soil 49: 711-715.
Allen, O.N. and Allen, E.K., 1981, "The Leguminoseae. A Source Book of Characteristic, Uses and Nodulation", MacMillan ed. ISBN: 0333-32221-5
Andersen, S.J. and Phillips, D.A., 1976, Effect of protein additives on acetylene reduction (nitrogen fixation) by Rhizobium in presence and absence of soybean cells, Plant Plysiol., 57:890-893.

Bednarski,M.A. and Reporter, M., 1978. Expression of rhizobial nitrogenase: influence of plant cell conditioned medium, Appl. Environ. Microbiol., 36:115-120.

Bergersen, F.J. and Turner, G.L., 1967. Nitrogen fixation by the bacteriod fraction of breis of soybean root nodules, Biochim. Biophys. Acta, 141: 507-515.

Bergersen, F.J., 1977. Nitrogenase in chemostat cultures of rhizobia, in: "Recent Developments in Nitrogen Fixation", W. Newton, J.R. Postage, C. Rodriguez-Barrueco, eds., New York, Academic Press, pp 309-320.

Child, J.J. and LaRue, T.A., 1974, A simple technique for the establishment of nitrogenase in soybean callus culture, Plant Physiol., 53:88-90.

DeMoranville, C.J., Kaminski, A.R., Barnett, N.M., Bottino, P.J., and Blevins, D.G., 1981, Substances from cultured soybean cells which stimulate or inhibit acetylene reduction by free-living Rhizobium japonicum, Physiol. Plant., 52:53-58.

Glenn, A.R., McKay, I.A., Arwas, R. and Dilworth, M.J., 1984, Sugar metabolism and the symbiotic properties of carbohydrate mutants of Rhizobium leguminosarum, J. Gen. Microbiol., 130:239-245.

Gotz, E.M., 1980. Attachment to plant root surface and nitrogenase activity associated with Petunia plants, Z. Pflanzenphysiol., 98:465-470.

Hardy, R.W.F., Holstein, R.D., Jackson, E.K., Burns, R.C., 1968, The acetylene-ethylene assay for N_2-fixation: laboratory and field evaluation, Plant Physiol, 43:1185-1207.

Keister, D.L., 1975, Acetylene reduction by pure cultures of rhizobia, J. Bacteriol., 123:1265-1268.

Kurz, W.G.W., and LaRue, T.A., 1975, Nitrogenase activity in rhizobia in absence of plant host, Nature, 256:407-409.

Martins-Loução, M.A. and Rodriguez-Barrueco, C., 1981, Establishment of proliferating callus from roots, hypocotyls and cotyledons of carob (Ceratonia siliqua L.) seedlings, Z. Pflanzenphysiol, 103:297-303.

Martins-Loução, M.A. and Rodriguez-Barrueco, C., 1982, Studies on nitrogenase activity of carob (Ceratonia siliqua L.) callus cultures associated with Rhizobium in: "Plant Tissue Culture", Fujiwara, ed., pp 671-672.

Martins-Loução, M.A. and Rodriguez-Barrueco, C., 1983, Ethylene production by carob (Ceratonia siliqua L.) callus cultures on varying media, Physiol. Plant, 58:204-208.

Martins-Loução, M.A., 1985, Estudios fisiológicos e microbiológicos da associação da alfarrobeira (Ceratonia siliqua L.) com bacterias de Rhizobiaceae, thesis, Lisboa.

McComb, J.A., Elliot, J. and Dilworth, M.J., 1975, Acetylene reduction by Rhizobium in pure culture, Nature, 256:409-410.

Pagan, J.D., Child, J.J., Scowcroft, W.R. and Gibson, H.H., 1975, Nitrogen fixation by Rhizobium cultured on a defined medium, Nature, 256: 406-407.

Scowcroft, W.R, and Gibson, A.H., 1975, Nitrogen fixation by Rhizobium associated with tobacco and cowpea cell cultures, Nature, 253:351-352.

Storey, R., Rainley, K., Pope, L. and Reporter, M., 1979, In vitro nitrogenase activity of Rhizobium japonicum affected by conditioned medium from culture plant cells, Plant Sci. Lett., 14:253-258.

Vincent, J.M., 1970, "A manual for the practical study of root nodule bacteria", IBP Handbook No. 15.

Werner, D., 1976, Nitrogenase activity in the in vitro symbiosis of Rhizobium japonicum and tissue cultures of Glycine max. and in Rhizobium in pure cultures on defined media, Ber. Dtsch. Bot. Ges., 89:563-574.

Wilcockson, J. and Werner, D., 1978, Nitrogenase activity by Rhizobium japonicum growing on agar surfaces in relation to slime production, growth and survival, J. Gen. Microbiol., 108:151-160.

ON THE EVOLUTION OF ALCOHOL TOLERANCE IN MICROORGANISMS

Lonnie O. Ingram and Kenneth M. Dombek

Department of Microbiology and Cell Science and Department
of Immunology and Medical Microbiology, 1052 McCarty Hall
University of Florida, Gainesville, Florida, 32611 U.S.A.

SUMMARY

   The accumulation of fermentation products such as ethanol in the
microbial environment represents a form of stress in many respects analogous
to extremes of pH, salt and temperature. Thus it is not surprising that
microorganisms have evolved with various levels of alcohol tolerance. In
this report, we have summarized the changes in membrane composition
associated with increased ethanol tolerance and have used these to formulate
a hypothesis for the primary action by which ethanol damages cells and
prevents cell growth.

   Microorganisms have evolved to live and grow in a wide range of
environmental extremes throughout the world. These include organisms such
as thermophiles, cryophiles, and barophiles which are adapted to physical
environmental extremes and organisms such as halophiles, alkalinophiles and
acidophiles which are adapted to chemical environmental extremes. In all
cases, organisms adapted to survival and growth in these extremes of
environment have evolved modifications in membrane structure, not found in
most organisms living in more moderate environments. Indeed, many of these
modifications such as the evolutionary replacement of ester linkages by
ether linkages which resist hydrolysis at extremes of pH are absolutely
essential to maintain an intact membrane structure (Brock, 1978; Kushner,
1978). The stability of the plasma membrane as an effective semipermeable
barrier is required for cellular growth and survival. It is this barrier
function of the plasma membrane which permits the compartmentalization and
concentration of molecules and processes which we term life. For survival
and growth in environmental extremes, evolutionary changes in soluble
proteins and enzymes such as increased thermotolerance would be without
consequence unless preceeded by evolutionary modifications which permit the
maintenance of an effective plasma membrane.

   The accumulation of organic fermentation products such as ethanol in
the microbial environment represents another type of chemical environmental
stress to which microorganisms have evolved different levels of tolerance
for growth. Figure 1 illustrates the sensitivity of three different
organisms to growth inhibition by ethanol. From this graph it is clear that
the most widely used experimental microorganism, Escherichia coli, is much
less ethanol tolerant than the obligately fermentative bacterium, Zymomonas
mobilis. The most commonly used organism for commercial alcohol production,

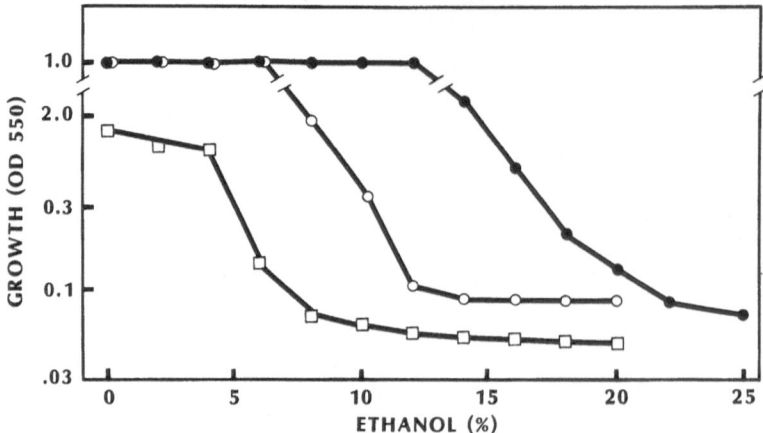

Fig. 1. Comparison of the sensitivity of different organisms to growth inhibition by ethanol. Growth was
measured as optical density at 550 nm after
incubation at 30°C for 48 hours. Symbols: □,
E. coli; O, Z. mobilis; ●, L. heterohiochii

Saccharomyces cerevisiae, is roughly equivalent to Z. mobilis in terms of
its ethanol tolerance for growth. The lactic acid group of bacteria have
long been known to possess increased ethanol tolerance and this criterion
has frequently been used for the taxonomic separation of this group. The
relatively high level of ethanol tolerance of the lactic acid bacteria is
perhaps not surprising when one considers that these organisms are
obligately fermentative and are specialized for successful competition in
fermentative environments. The most alcohol tolerant vegetative organisms
known are found among the lactic acid group of bacteria. These are the two
common spoilage organisms of sake', Lactobacillus heterohiochii and
L. homohiochii (not shown).

During the past few years, our laboratory and others have investigated
the relationship between membrane composition and alcohol tolerance in
different microorganisms. Our studies initially focussed on E. coli, an
organism which normally produces only small amounts of ethanol as a
fermentation product (Chesbro et al., 1979). This organism is highly
regulated and adaptable to both oxidative and fermentative growth. The
addition of ethanol to growing cells of E. coli resulted in a number of
changes in membrane composition which began immediately and were reversible
upon ethanol removal.

Figure 2 illustrates the effects of growth in the presence of ethanol
on membrane fatty acid composition. With increasing concentrations of
ethanol, the proportion of vaccenic acid (18:1) increased dramatically at
the expense of palmitic acid (16:0) while palmitoleic acid (16:1) remained
essentially constant (Ingram, 1976; Ingram et al., 1980a). This change is
analogous to the changes in fatty acid composition which occur in response
to lowering the growth temperature of E. coli (Marr and Ingraham, 1962).
The ethanol-induced changes in fatty acid composition occurred in all major
classes of phospholipids (Berger et al., 1980). This suggested that ethanol
affected cellular fatty acid composition by acting during the biosynthesis
of the common precursor, phosphatidic acid. The types of changes induced by
ethanol were also induced by growth in the presence of a wide variety of
other small, non-ionic organic molecules which were relatively polar
indicating the lack of a strict structural requirement but rather a more
generalized mechnanism of action (Ingram, 1977a).

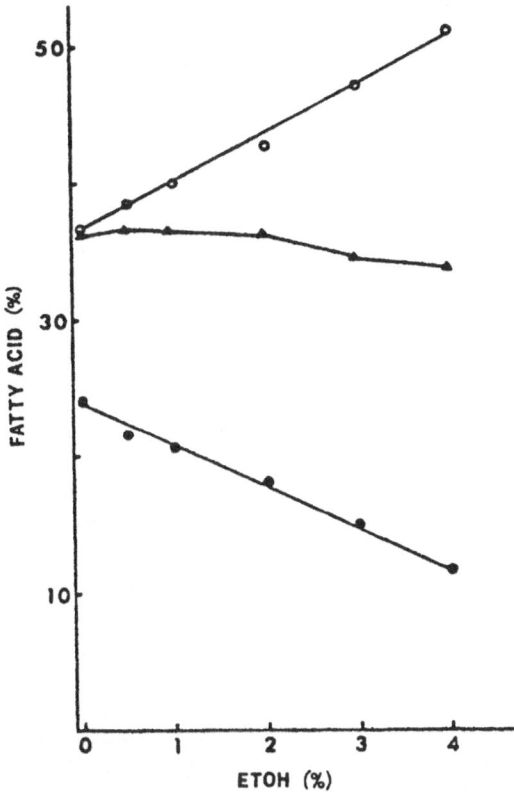

Fig. 2. Effects of ethanol on the fatty acid
composition of E. coli. Cells were
grown at 30°C in the presence of different concentrations of ethanol. Fatty
acid compositions were determined
following lipid extraction and trans-
esterification by gas liquid chromatography. Symbols: O, vaccenic acid
(18:1); ▲, palmitoleic acid (16:1);
●, palmitic acid (16:0).

The mechanism by which ethanol induced changes in fatty acid
composition was examined in some detail. Studies with mutants indicated
that ethanol exerted its principal effect on fatty acid composition at the
level of fatty acid biosynthesis, altering the types of fatty acids which
were synthesized and available for assembly into phospholipids (Buttke and
Ingram, 1978). This direct effect of ethanol on the soluble enzymes of
fatty acid biosynthesis was subsequently confirmed during in vitro studies
(Buttke and Ingram, 1980). Using mutants defective in several enzymes of
fatty acid biosynthesis, the β-ketoacyl ACP synthetase II enzyme was
indentified as the most likely target for this ethanol effect and for the
temperature-induced changes in fatty acid composition.

Subsequent studies provided a clue to the common mechanism by which a
decrease in temperature and the addition of ethanol could affect fundamental
cellular processes. Both of these treatments alter the structure of water
and decrease the strength of hydrophobic interactions (Ingram and Vreeland,
1980; Ingram, 1981). This led to the prediciton and subsequent confir-
mation in vivo and in vitro that chaotropic salts and other compounds which

weaken the strength of hydrophobic interactions have a similar effect to that of ethanol on membrane fatty acid composition and on fatty acid biosynthesis in vitro (Ingram, 1982). Although the precise mechanism has yet to be defined, it is likely that these permeable compounds cause changes in the colligative properties of the cytoplasm which alter the way in which the acyl-acyl carrier protein (acyl-ACP) substrates of fatty acid biosynthesis bind hydrophobic regions of β-ketoacyl ACP synthetase II.

Growth of E. coli in the presence of ethanol also caused changes in phospholipid composition (Ingram, 1977b). Increasing concentrations of ethanol resulted in a decrease in the proportion of phosphatidylethanol-amine (the most abundant phospholipid) with increases in the proportion of the acidic phospholipids, cardiolipin and phosphatidylglycerol. This resulted from the preferential inhibition of phosphatidylserine synthetase and led to the production of lipid-poor membranes which exhibited a dramatic reduction in the phospholipid to protein ratio (Dombek and Ingram, 1984).

Having characterized ethanol-induced changes in membrane composition, the next question of major interest was the significance of these changes. Mutants of E. coli which are defective in fatty acid biosynthesis were used to alter fatty acid composition at will by adding fatty acid supplements. Using these mutants, we were able to demonstrate that an increase in vaccenic acid (18:1) was beneficial for growth and survival in the presence of ethanol (Ingram et al., 1980). In unpublished studies, we found that trans-fatty acids were even more beneficial than cis-fatty acid indicating that the increase in acyl chain length rather than an increase in the unsaturated fatty acid content was important. Doubly unsaturated lipids were not helpful and cells enriched in palmitic acid (16:0) were hyper-sensitive to inhibition by ethanol (Ingram et al., 1980).

Other studies by Clarke and Beard (1979) examining alcohol-resistant and alcohol-sensitive mutants of E. coli provided evidence that increased abundance of acidic phospholipids was beneficial for growth in the presence of ethanol. Recent studies concerning the effects of ethanol on membrane organization have provided evidence that the ethanol-induced decrease in lipid/protein ratio may also be beneficial to a point, compensating for the fluidization of the membrane by ethanol and decreasing the membrane area available for the leakage of intracellular constitutents (Dombek and Ingram, 1984).

In some respects, S. cerevisiae, is similar to E. coli in that both are highly adaptable organisms which are capable of oxidative and fermentative metabolism. Although these organisms have extremely different enzymatic machinery for fatty acid biosynthesis (Fulco, 1974), the fatty acid composition of S. cerevisiae is very similar to that of E. coli but with oleic acid (18:1) rather than vaccenic acid (18:1). Studies by Beaven et al. (1982) have demonstrated that the growth of S. cerevisiae in the presence of ethanol results in changes in fatty acid composition completely analogous to those which occur in E. coli with the replacement of saturated fatty acids (primarily 16:0) with a longer chain mono-unsaturated fatty acid, oleic acid (18:1). This may be interpreted as an example of convergent evolution involving completely different enzymatic systems for the biosynthesis of 18:1 fatty acids (Figure 3). Other studies with S. cerevisiae from Dr. Rose's laboratory have shown that cell sensitivity to killing by ethanol can be modified by changing either the fatty acid or sterol composition of the yeast membrane (Thomas et al., 1978). These results are consistent with the hypothesis that the ethanol-induced changes in fatty acid composition are beneficial in yeast as they appear to be in E. coli.

YEAST (O$_2$ –dependent)

acetate malonate $\longrightarrow$ FAS $\longrightarrow$ 16:0 $\longrightarrow$ 18:0

oxygen

16:1   18:1

E. coli (Anaerobic)

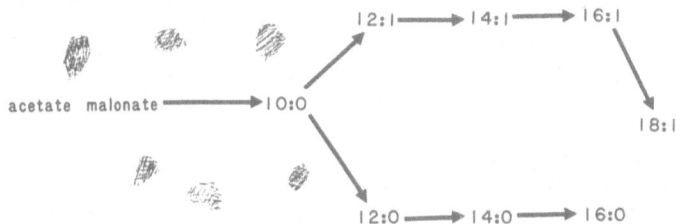

acetate malonate $\longrightarrow$ 10:0

12:1 $\longrightarrow$ 14:1 $\longrightarrow$ 16:1

18:1

12:0 $\longrightarrow$ 14:0 $\longrightarrow$ 16:0

Fig. 3. Comparison of biosynthetic routes for fatty acids in S. cerevisiae (upper) and E. coli, Z. mobilis and Lactobacilli (lower). Yeast synthesize fatty acids via a multifunctional fatty acid synthetase system and insert double bonds into homologous saturated fatty acids using an oxygen dependent desaturase system. E. coli and many other bacteria utilize soluble individual enzymes for fatty acid synthesis and insert double bonds using a non-oxygen-requiring pathway during the elongation of the fatty acid chains.

Z. mobilis is a gram negative bacterium which produces ethanol and carbon dioxide as major fermentation products. This organism does not undergo dramatic changes in fatty acid composition in response to added ethanol or during the fermentative accumulation of ethanol (Carey and Ingram, 1983). However, over 70% of the acyl chains in the phospholipids of this organism are 18:1, vaccenic acid. This organism is highly specialized for obligate ethanol production and exhibits an extreme of the inducible trends observed with both E. coli and S. cerevisiae. Other changes which are induced by ethanol in Z. mobilis include the synthesis of lipid-poor membranes and increases in the proportion of the terminal products of phospholipid biosynthesis, cardiolipin and phosphatidylcholine.

Recent studies by Bringer et al. (1985) have revealed a new ethanol-inducible change in the membranes of Z. mobilis, a dramatic increase in the hopinoid content. Hopinoids are thought to be evolutionary precursors of sterols and are found in a variety of bacteria. Increases in hopinoids would be expected to increase the barrier properties of the plasma membrane and decrease the passive permeability of polar solute molecules.

L. heterohiochii and L. homohiochii are the two most alcohol tolerant organisms known (Kitahara et al., 1957a & b; Demain et al., 1961). These organisms have very specialized membranes and appear to have adopted different evolutionary routes for membrane modification and ethanol tolerance (Uchida and Mogi, 1973). L. homohiochii exhibits an ethanol inducible increase in the proportion of vaccenic acid (18:1) such that less than 3% of its membrane acyl chains are saturated fatty acids (primarily 16:0 and 18:0) (Uchida, 1975). In contrast, L. heterohiochii also exhibits an ethanol inducible increase in the proportion of unusually long chain mono-unsaturated fatty acids which include 18:1, 20:1, 22:1, 24:1, 26:1, 28:1 and 30:1 (Uchida, 1974a & b). Indeed, fatty acyl chains of 20 carbons and over constitute over 30% of the total fatty acids, an occurence not found elsewhere in the biological world. Further, L. heterohiochii actually requires moderate levels of ethanol for optimal growth (Demain et al., 1961). This ethanol-dependence is analogous to the stress-dependence which has accompanied the evolution of other microorganisms able to colonize environmental extremes (Brock, 1978; Kushner, 1978).

A common trend can be inferred from this comparison of the evolutionary changes in membrane structure which are associated with ethanol tolerance among different microorganisms. The single trend observed in all cases is an increase in acyl chain length and in the proportion of mono-unsaturated fatty acids. Since trans-fatty acids are equally effective in some cases, the increase in acyl chain length rather than the increase in unsaturation appears to be the beneficial aspect of this change in fatty acid composition. One major effect of an increase in mean acyl chain length is to increase the thickness of the hydrophobic core of the cell membrane, the cell's primary permeability barrier. Along these lines, ethanol is known to weaken the barrier function of the membrane and to increase cellular leakage (Dombek et al., 1985; Eaton et al., 1982). Increasing the thickness of the hydrophobic core would tend to offset ethanol damage to the barrier function of the membrane. At first glance, one might expect that saturated acyl chains rather than mono-unsaturates would be of even more benefit. However, most prokaryotic acyl transferase enzymes will not readily utilize saturated acyl chains longer than 16 carbons for phospholipid biosynthesis (Cronan, 1978). Further, the production of phospholipids with two saturated fatty acyl chains even of only 16 carbons in length is detrimental leading to phase separations within the membrane and increased membrane leakage (Ingram et al. 1982; McElhaney, 1984). This problem of phase separations is even more severe for longer chain, disaturated phospholipids.

Based upon a consideration of the changes in membrane lipid composition associated with increased alcohol tolerance for growth, we propose that the principal ethanol-damage to which cells have evolved resistance is the ethanol-induced increase in membrane leakage. The observed fatty acid changes are most consistent with this hypothesis and the other changes in membrane structure such as an increase in hopinoids (increase in the hydro-phobicity of the membrane core), a decrease in the proportion of phospho-lipids (decrease in the lipid surface area of the membrane available for leakage) and the retention of acidic phospholipids (lipids which bind proteins tightly and are most frequently required for maintenance of enzymatic function) can also be rationalized as contributing toward a reduction in this type of ethanol damage.

ACKNOWLEDGEMENTS

This work was supported in part by the National Science Foundation (DMB 8204928) and by the Florida Agricultural Experiment Station (publication number 6356).

REFERENCES

Beaven, M.J., Charpentier, C. and Rose, A.H., 1982, Production and tolerance in ethanol in relation to phospholipid fatty acyl composition in Saccharomyces cerevisiae, NCYC 431, J. Gen. Microbiol., 128:1447-1455.

Berger, B., Carty, C.E. and Ingram, L.O., 1980, Alcohol-induced changes in phospholipid molecular species of Escherichia coli, J. Bacteriol., 142:1040-1048.

Bringer, S., Hartner, T., Poralia, K. and Sahm, H., 1985, Influence of ethanol on the hopinoid content and fatty acid pattern in batch and continuous culture of Zymomonas mobilis, Arch. Microbiol., 140:312-316.

Brock, T.D., 1978, Thermophilic Microorganisms and Life at High Temperatures, Springer-Verlag, New York.

Buttke, T.M. and Ingram, L.O., 1980, Ethanol-induced changes in lipid composition of Escherichia coli: Inhibition of saturated fatty acid synthesis in vitro, Arch. Biochem. Biophys., 203:465-471.

Buttke, T.M. and Ingram, L.O., 1978, Mechanism of ethanol-induced changes in lipid composition of Escherichia coli: Inhibition of saturated fatty acid biosynthesis in vivo, Biochemistry, 17:637-644.

Carey, V.C. and Ingram, L.O., 1983, Lipid composition of Zymomonas mobilis: Effects of ethanol and glucose, J. Bacteriol., 154:1291-1300.

Chesbro, W., Evans, T. and Eifert, R., 1979, Very slow growth of Escherichia coli, J. Bacteriol., 139:625-638.

Clark, D.P. and Beard, J.P., 1979, Altered phospholipid composition in mutants of Escherichia coli sensitive or resistant to organic solvents, J. Gen. Microbiol., 113:267-274.

Dombek, K.M., Holt, A.S. and Ingram, L.O., 1985, Effects of temperature on the potency of ethanol as an inhibitor of growth and membrane function in Zymomonas mobilis, Dev. Industrial. Microbiol., 26: in press.

Cronan, J.E., Jr., 1978, Physical properties of membrane lipids: Biological relevance and regulation. Ann. Rev. Biochem., 47:163-189.

Demain, A.L., Rickes, E.L., Henlin, D. and Barnes, E.C., 1961, Nutritional studies on Lactobacillus heterohiochii, J. Bacteriol., 81:147-153.

Dombek, K.M. and Ingram, L.O., 1984, Effects of ethanol on the Escherichia coli plasma membrane, J. Bacteriol., 157:233-239.

Eaton, L.V.C., Tedder, T.F. and Ingram, L.O., 1982, Effects of fatty acid composition on the sensitivity of membrane functions to ethanol, Substance and Alcohol Actions/Misuse, 3:77-87.

Fulco, A.J., 1974, Metabolic alterations of fatty acids, Ann. Rev. Biochem., 43:215-241.

Ingram, L.O., 1982, On the regulation of fatty acid composition in Escherichia coli: A proposed common mechanism for changes induced by ethanol, chaotropic agents and a reduction in growth temperature, J. Bacteriol., 149:166-172.

Ingram, L.O., 1981, Mechanism of lysis of E. coli by ethanol and other chaotropic agents, J. Bacteriol., 146:331-336.

Ingram, L.O., 1977a, Changes in lipid composition of Escherichia coli resulting from growth with organic solvents and food additives, Appl. Environ. Microbiol., 33:1233-1236.

Ingram, L.O., 1977b, Preferential inhibition of phosphatidyl ethanolamine synthesis in Escherichia coli by alcohols, Can. J. Microbiol., 23:779-789.

Ingram, L.O., 1976, Adaptation of membrane lipids to alcohols, J. Bacteriol., 125:670-678.

Ingram, L.O. and Buttke, T.M., 1984, Effects of alcohols on micro-organisms, Adv. Microbial. Physiol., 25:253-300.

Ingram, L.O., Dickens, B.F. and Buttke, T.M., 1980a, Reversible effects of ethanol on Escherichia coli, Adv. Exp. Med. Biol., 126:299-338.

Ingram, L.O., Eaton, L.C., Erdos, G.W. and Tedder, T.F., 1982, Unsaturated fatty acid requirement in Escherichia coli: Mechanisms of palmitate-induced inhibition of growth of strain WNI, J. Membrane Biol., 65:31-40.

Ingram, L.O. and Vreeland, N.S., 1980, Differential effects of ethanol and hexanol on the <u>Escherichia coli</u> cell envelope, J. Bacteriol., 144:481-488.

Ingram, L.O., Vreeland, N.S. and Eaton, L.C., 1980b, Alcohol tolerance in <u>Escherichia coli</u>, Pharmacol. Biochem. Behav., 13:191-195.

Kitahara, K., Kaneko, T. and Goto, O., 1957, Taxonomic studies on the hiochi-bacteria, specific saprophytes of sake'. I. Isolation and grouping of bacterial strains, J. Gen. Appl. Microbiol., 3:102-110.

Kitahara, K., Kaneko, T. and Goto, O., 1957, Taxonomic studies on the hiochi-bacteria, specific saprophytes of sake'. II. Identification of hiochii-bacteria, J. Gen. Appl. Microbiol., 3:111-120.

Kushner, D.J. (ed.), 1978, Microbial Life in Extreme Environments, Academic Press, New York.

Marr, A.G. and Ingraham, J.L., 1962, Effect of temperature on the fatty acid composition of <u>Escherichia coli</u>, J. Bacteriol., 84:1260-1267.

McElhaney, R.N., 1984, The relationship between membrane lipid fluidity and phase state and the ability of bacteria and mycoplasmas to grow and survive at various temperatures, <u>in</u>: Kates, M. and Mason, L.A. (eds.), "Membrane Fluidity", Plenum Publishing Corporation, New York.

Thomas, D.S., Hossack, J.A. and Rose, A.H., 1978, Plasma membrane lipid composition and ethanol tolerance in <u>Saccharomyces cerevisiae</u>, Arch. Microbiol., 117:239-245.

Uchida, K., 1975, Alteration of the unsaturated to saturated ratio of fatty acids in bacterial lipids by alcohols, Agr. Biol. Chem., 39:1515-1516.

Uchida, K., 1974a, Occurrence of saturated and mono-unsaturated fatty acids with unusually-long-chains (C_{20}-C_{30}) in <u>Lactobacillus heterohiochii</u>, an alcoholophilic bacterium, Biochem. Biophys. Acta 348:86-93.

Uchida, K., 1974b, Lipids of alcoholophilic lactobacilli. II. Occurrence of polar lipids with unusually long acyl chains in <u>Lactobacillus heterohiochii</u>, Biochim. Biophys. Acta 369:146-155.

Uchida, K. and Mogi, K., 1973, Cellular fatty acid spectra of hiochii bacteria, alcohol-tolerant lactobacilli and their group separation, J. Gen. Appl. Microbiol., 19:233-249.

INDUSTRIAL APPLICATIONS OF PLANT CELL CULTURES: METABOLITE PRODUCTION

Vincent Petiard[1] and Paul Steck[2]

[1] L.E.R. Synthelabo, Tours, France
[2] Sanofi/Elf Bio Recherche, Toulouse, France

INTRODUCTION

In vitro plant cell and tissue culture has two major fields of application.

Agronomical applications consist of:

- either the rapid multiplication of a selected plant by the multiple production of plants identical to the original plant,

- or exploitation of the variability found in vitro (or which may be induced using in vitro methods) in order to obtain a new plant, different from that originally used and more efficient according to certain defined criteria.

Industrial applications, the subject of the present article, likewise present two aspects parallel to those described above:

- the production of a known molecule, using the biosynthesis capacities of plant cells bred in a bioreactor,

- an innovative aspect, exploiting the new source of variability accessible in vitro to obtain new molecules.

It is not necessary here to dwell on the number and diversity of the products of plant origin which we use every day. We will not enter into the agro-alimentary domain, which has been discussed elsewhere, nor that of material or energy production, for which plant biotechnology is still much too expensive: for the moment, we need only be concerned with the synthesis of compounds which are rare, onerous, or difficult to obtain by other channels. This explains why, in the first place, it is the pharmaceutical industry, and perhaps the flavouring and perfume industries, which will benefit from products obtained using plant biotechnology.

Each time we present a prescription at the chemist's, there is more than one chance in two that the medicine we will receive has originated from natural substances. The market for plant-originating pharmaceutical products of plant origin in developed countries may therefore be estimated at more than 20 billion dollars.

With regard to the production of certain of these substances by plant cell culture, it should be noted that, due to the present high cost of such technology, only a few rare or extremely onerous products may have a claim to _in vitro_ biosynthesis production for the time being.

Let us now recapitulate the various ways of obtaining substances of plant origin. Classically speaking, useful substances are extracted from plants either gathered from the wild or purpose-cultivated. Hemisynthesis has often been used to obtain more active or less toxic substances by slight modification of the chemical structure of plant-extracted molecules. It also allows a less expensive or more readily available raw material to be exploited by the use of an intermediary extracted from another species (e.g.tabersonin from _Voacanga_). Chemical synthesis has often taken inspiration for the structure of a medicine from natural models. However, chemistry does not always manage to reproduce viable natural substances, producing, for example, mixtures of isomers, where the plant only produces the single active isomer.

The choice of production method is, of course, dictated by precise economic criteria depending on the size of the market, cost, the availability of raw materials, import possibilities and therefore the balance of trade, and complex socio-economic factors in the case of narcotics such as alkaloids extracted from the poppy, for example.

Plant biotechnology proposes two new approaches (Table 1):

- _in vitro_ production of the active substance by cell culture, as described below,

- intervention of cultivatable medicinal species at the agronomical level.

It also opens the way to innovation, as already mentioned, making possible the discovery (and subsequent production) of new active substances.

This possibility of improved production of a natural substance in the whole plant should be kept in mind. The generally restricted size of the market of a given medicinal plant does not generally justify the setting-up

Table 1. Various Ways of Obtaining Substances of Plant Origin

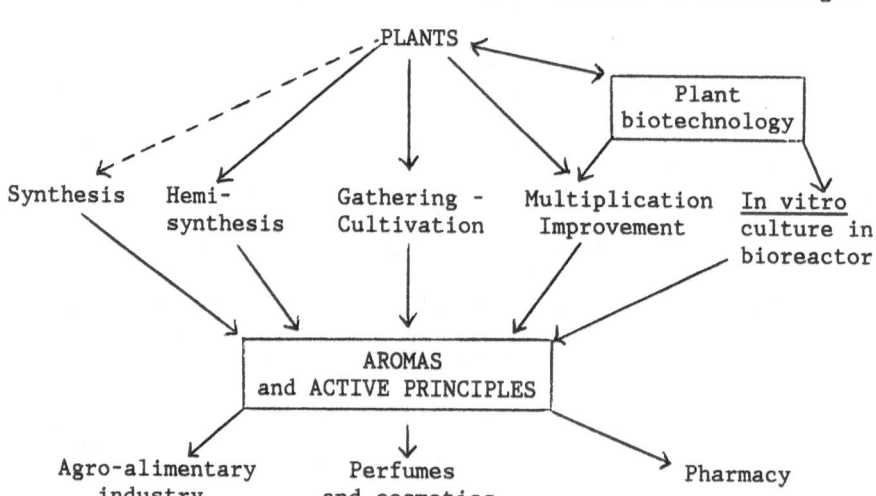

140

of a selection programme along classical lines. The natural variability of certain medicinal species is sometimes vast, as in the case studied by Zenk et al., (1977) on the alkaloid content of Catharanthus roots of varying origin. This variability makes certain medicinal species excellent candidates for micropropagation: rich individuals are selected for cloning by vegetative multiplication to supply a crop in the field, or, if the process is too heavy on manpower, clone hybrid-based seeds will be produced. In this way, the alkaloid content of Catharanthus plants has been easily doubled (Abou-Mandour et al., 1979). These methods have already been applied to various aromatic (e.g. for drinks, fennel for example (Dumanoir et al., 1985)) or medicinal plants (Table 2).

METHODOLOGICAL STAGES

From Plant to Bioreactor

Table 3 shows the usual development stages in the production process of an active substance of plant origin. This may be narrowed down to three key stages:

- development of the productive strains,

- mass culture,

- extraction of the active substances.

For each of these points we shall look at the present state of the technology.

At this point, a possible preliminary stage should be noted: normal selection of the wild plant, allowing highly productive plants of the substance to be obtained in vitro to be used to establish the primary cultures.

Zenk observed a degree of variability with regard to the serpentine content of Catharanthus roseus roots, and was able to establish in many cases a correlation between the active substance content of tissue cultures and the productivity of the mother plant from which these cultures

Table 2. In Vitro Vegetative Multiplication of Medicinal Plants

Some examples

Pinellia ternata	Anti-emetic	Shoyama, Y.	1983
Digitalis lanata	Cardiotonic	Erdei, I.	1981
Coptis japonica	Berberine	Shono, K.	1972
Datura innoxia	Tropanics	Hiraoka, N.	1974
Hyoscyamus scopolia	Tropanics	Shimomura, K.	1980
Panax ginseng	Ginsengosides	Chang, N.C.	1980
Rhemannia glutinosa	(Chinese drug)	Shoyama, Y.	1983
Discorea deltoidea	Diosgenin	Staba, E.J.	1969

Table 3. Methodological Stages

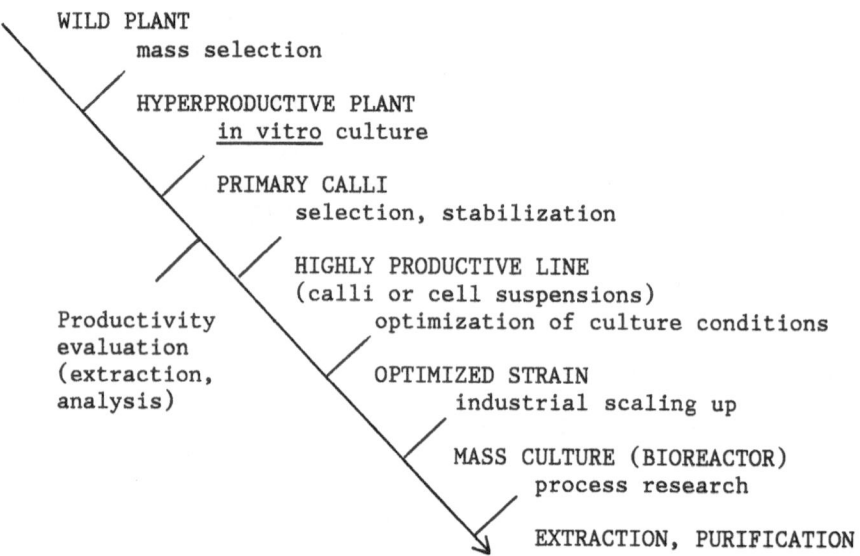

WILD PLANT
    mass selection

HYPERPRODUCTIVE PLANT
    in vitro culture

PRIMARY CALLI
    selection, stabilization

HIGHLY PRODUCTIVE LINE
(calli or cell suspensions)
    optimization of culture conditions

OPTIMIZED STRAIN
    industrial scaling up

MASS CULTURE (BIOREACTOR)
    process research

EXTRACTION, PURIFICATION

Productivity
evaluation
(extraction,
analysis)

originated. Although no generalisation can be made on this matter, the pre-selection of hyperproductive individuals may therefore be a useful initiative.

Another stage of decisive importance in the perfection of a production process is that known as "downstream processing", which is concerned with the extraction and purification on an industrial scale of the active material produced.

It is clear that, at every stage, the work is conditioned by the capacity to correctly evaluate productivity, i.e. to know how to extract, analyze, and quantify metabolites produced. We shall see that extraction may pose certain problems specific to in vitro culture. Rapid, precise, and above all sensitive dosage is often crucial. RIA dosage, for example, allows a greater degree of sensitivity than HPLC, whereas an ELISA-type enzymo-immunological technique will generally be less cumbersome and more rapid than RIA.

THE STATE OF THE TECHNOLOGY - KEY PROBLEMS

Development of the Productive Strains

What happens when a plant cell is cultivated in vitro?

Within the plant itself, the cell receives a complex ensemble of signals from the surrounding tissue, the other organs of the plant, from the environment and external conditions, signals which will engage the cell in a certain form of differentiation, or specialization. It is obliged to perform a particular expression programme, acquiring a well-defined structure and function at the expense of its other potentialities. Like all beings living with a "society", it takes on its own "personality", in the form of a certain cell type - radicular, caulinary, follinary - performing a certain type of biosynthesis, storage, transport, etc., role. It manifests and develops some of its characteristic properties to the detriment of the remaining possibilities of expression written into its genetic programme.

Once dissociated from the organism from which it originated and put into in vitro culture, the cell is freed from the characteristic constraints of its in vivo environment. Tissue cultures and cell suspensions are often described as "de-differentiated", meaning that the cells of which they are comprised have lost the particular characteristics they showed within the whole organism. By "de-repression" of their genetic programme, they have rediscovered a kind of "topipotency", or "new youth".

In this situation it will be possible to re-direct them towards new types of differentiation induced by the environments to which they are exposed, thus causing them to express new properties. In order to obtain a productive strain, this spontaneous variability must first of all be exploited, or even induced if necessary. The next step, on the contrary, entails the elimination of this variability to stabilize the strain, thus conserving its productivity characteristics for exploitation.

Variability is sometimes expressed by readily visible characteristics: on colourless tissue, variant cells capable of anthocyan synthesis may appear. This island of cells may be cultured out separately, and a strain conserving this new property may be isolated. Strains of Catharanthus obtained by such somaclonal variations are today well established at Tours, where each strain keeps its own specific colour, i.e. its particular pigmentary metabolism, in spite of the fact that all the strains originated from the same initial culture. Other morphological variations may be observed, such as callus consistency (more or less compact, more or less friable), which may be a useful characteristic for easier passge of the strain into a liquid medium.

The production variability of a desired metabolite will of course rarely be expressed by such readily visible characteristics. Various conditions will have to be tried to reveal the extent of the biosynthesis capacity of a given product. By varying the quantity of alkaloids produced by a great number of Catharanthus roseus calli, Zenk et al. (1977) showed that, if a high proportion of clones contained relatively little Serpentine or Ajmalicine, there was a certain probability of finding clones producing 5 or 10 times more than the average, thus opening the way to the selection of hyperproductive clones.

Variability may also be expressed by very diverse capacities for the bioconverstion of a given substrate. If, for example, the capacity required is that of glycosylation or hydroxylation, a strain will be sought capable of performing this specific reaction with a high yield on a particular substrate.

The very existence of this biochemical variability of tissue cultures has only recently been recognized. Until 1976-78, it was assumed that certain structures or specialized organs like those existing in the whole plant were indispensable for the biosynthesis of the secondary metabolites apparently associated with these structures in situ: in other words, that certain biosynthetic operations would be impossible in vitro as long as the morphogenesis was repressed. Until two or three years ago, it was still widely thought that, at least for substances found in particular tissues in the whole plant, their production by morphologically undifferentiated cells could not be hoped for. Today the problem seems definitively removed: secondary products can occur in the absence of any differentiation, e.g. the alkaloids stored in the laticiferous cells of the poppy on one hand, and volatile compounds, such as the essential oils accumulated in the secretory glands, for example. This biosynthesis occurs with a very high degree of variability according to the strains considered, the majority of which produce very few or no secondary metabolites. However, the fundamental interest of these variability results lies in the fact that it is possible, by cloning (real cloning or on small aggregates), to select strains producing as much as, and even more than the whole plant. Table 4 gives a few examples.

These comparisons, based on concentration in relation to dry weight, are virtually meaningless if the production time of plant matter is not taken into account. Let us compare, for example, the productivity of a cell suspension of Coptis japonica (13.2% of berberine in a three-week old culture, i.e. 44 mg/g per week), to that of the root of the same plant (3% in five-year old roots, i.e. 83 μg/g per week): berberine is produced 530 times more quickly in vitro than in vivo. In the case of Lithospermum erythrorhizon, shikonin (12%, i.e. 60 mg/g per week in a two-week old cell suspension, and 1.5%, i.e. 48 μg/gper week in six-year old roots) is produced 1250 times more quickly in vitro than in the material from which it is traditionally extracted. These two comparisons, although approximate as it is clear that production is not continuous in these two types of material, clearly show that a cell may become adapted in vitro for a particular biosynthetic task.

Spontaneous in vitro variability may also be expressed qualitatively. A strain of Catharanthus roseus (Petiard, 1982), for example, allowed the isolation and identification of 13 alkaloids, of which 6 were new, i.e. had hitherto never been isolated from this species, 3 of these 6 had not been described in other plants, and one of which was totally unknown, even by chemical synthesis. The transitory de-stabilization of cell metabolism during its passage in vitro thus opens new paths for biosynthesis. Above all, this illustrates the interest in in vitro culture with regard to the discovery of new compounds, the industrial aptness of which may be more easily defended, if required.

In the exploitation of in vitro spontaneous variability to obtain productive strains, it is clear that a slight disturbance in the culture conditions may easily induce a metabolic variation. This brings us to the problem of subsequent strain stabilization: strains should be conserved in extremely constant conditions, and subcultures should be perfectly standarized for all factors within our control.

But before stabilization, if the diversity of the initial material is to be increased even more, the following may be envisaged:

- mutagenesis on diploid, or better still, haploid cells,

Table 4. A Few Recent Examples of in vitro-Exceeding in vivo Content

Compound	Species	Content (in vitro/in vivo)	References
Vomilenine	Rauwolfia serpentina	x 51	Stockigt, 1981
Jatrorrhizine	Dioscoreophyllum cumminsii	x 100	Furuya, 1983
Caffein	Coffea arabica	x 92	Waller, 1983
Berberine	Coptis japonica	x 6	Yamada, 1983
Tripdiolide	Trypterygium wilfordii	x 4.3	Misawa, 1983

- somatic fusion between protoplasts of different species, as performed by Yamada to find a hybrid metabolism between Euphorbia milii producing anthocyans, and Coptis japonica synthesizing berberine,

- photoautotrophy, for improved biosynthesis dependent upon functional chloroplasts,

- plant cell transformation, by the integration of coding exogenous DNA, for example, for a limiting enzymatic stage (in as much as it is known). An example of this was given by Berlin et al. (1983), who tried to introduce into lupin cells a coding bacterial gene for lysine-decarboxylase, a key enzyme in the biosynthesis of quinolizidines. A further example is that of the Monsanto Company, who attempted the transfer of the chorionic gonadotropin gene into Petunia cells.

An easy means of giving a strain extraneous information is to use as substrate a molecule from another species or which has been chemically synthesized. Its ability to perform the most varied bioconversions will thus be tested (e.g. see Umetami et al., 1982).

These various ways of inducing supplementary variability are still little used for in vitro metabolite production. But in the future they may facilitate access to certain hyperproductive strains and also, and above all, increase the chances of discovering new molecules.

What are the possibilities, then, of stabilizing a strain with regard to certain characteristics (in particular a capacity for biosynthesis) to be able to use it industrially? This stabilization is possible on condition that, as we have already seen, culture conditions are kept rigourously constant during successive subculturing. A strain of Catharanthus, for example, has not lost its aptitude to produce alkaloids after four years of sub-culturing (Petiard and Courtois, 1984).

The problem of stabilization tends, however, to be solved in other ways. Various strain conservation methods are on the point of being perfected, such as hypoxia under oil (see J.M. Augereau's wallchart), culture at low temperature (up to 4°C) allowing sub-culturing frequency to be considerably reduced, and cryopreservation, which, as described by Seitz, has made a lot of progress in the last few years. The crux of the problem lies in the need for strains to conserve their viability and former characteristics with regard to growth and biosynthesis.

Mass Culture

The development of a plant cell production process supposes that the problem of their culture on a large scale, according to processes more or less resembling those of micro-organism fermentation, has been solved.

Plant cells multiply much more slowly (about 100 times slower) than bacteria. Even for populations in rapid growth periods, the time required to double the number of cells is often betwen 20 and 40 hours. This slow multiplication is obviously linked to the size of plant cells, which are about 200,000 times larger than bacteria.

Within the plant itself, in certain meristems, or in the formation of a somatic embryo, doubling times of between 6 and 8 hours have been observed. It is therefore probable that culture performances will be improved. The present record is for a doubling time of 15 hours obtained for a Nicotiana tabacum suspension at a scale of 20 m^3 by Noguchi et al. (1977), with the dry weight yield reaching 5 to 10 g/l per day.

It should be noted that the slow growth rate is not necessarily an insurmountable handicap, as it is not biomass that is generally sought, but the use of this biomass for its biosynthesis capacity. Guern (1979) showed that instantaneous enzymatic activity of plant biomass is on a par with that of bacteria, for example for polysaccharides of the cell wall. Moreover, plant biomass may be used over a much longer period, up to several months, as we shall see with immobilized cell systems.

The maximum density of cell suspensions may attain 25 to 30 g/l of dry weight, depending on the carbon source, and conversion yields may reach 75 to 85%, whereas for bacteria they often remain below 50%.

All types of culture system have been applied to plant cells: batch, semi-continuous, continuous, cell immobilization. All types of reactor may likewise be used for plant cells. In the past it has been pointed out that, due to their size, plant cells are more sensitive to shearing forces than bacteria, and for this reason air-lift stirring systems are sometimes preferred to impeller stirring systems. This would no longer appear to be a problem, as numerous impeller systems have been brought out, including the 20 m^3 fermenter of Noguchi. The problem of long-term sterility of plant cells in the face of rapid proliferation of bacteria has also been raised. Improved experimental sterility conditions have nowadays eliminated this type of problem.

Do the special characteristics of plant cells allow culture volumes as large as for bacteria? It is true that, with the 20 m^3 of Noguchi, we are still relatively far behind, but looking at the evolution of plant culture volumes over the last 20 years (Table 5), it seems likely that much greater volumes will be possible if required.

The domination of large-scale systems obviously requires the measurement and regulation of numerous culture parameters. As far as process automation is concerned, mass plant culture follows in the footsteps of industrial microbiology.

To sum up this second key question, it is nowadays possible to cultivate plant cell strains producing useful substances in various types of bioreactor and on an industrial scale.

Table 5. Development of Mass Culture of Plant Cells

Year	Volume (litres)	Species	Authors
1959	9	Gingko, Ilex	Tulecke and Nickel
1960	30-134	Lolium, Rosa	Tulecke and Nickel
1972	130-600	Nicotiana	Yasuda
1974	300	Petroselinium	Hahlbrock
1975-76	65-1500	Nicotiana	Kato
1977-82	20,000	Nicotiana	Noguchi, Hashimoto
1982	20,000	Nicotiana	Hashimoto

Source: Zenk (1982)

Extraction of the Active Substance

In view of the low growth rate of plant cells in culture, it is generally speaking more economically viable to extract the product from the culture medium rather than sacrifice the biomass.

As we have already seen, this biomass may survive for a long time, allowing biosynthesis capacities to be exploited for weeks in continuous systems, and for months in immobilized systems, for as long as the product is diffused into the medium.

Since plant secondary metabolites generally accumulate in the vacuoles within the cells, research into the factors favourizing the diffusion of products and their excretion into the culture medium is of primary importance.

The percentage of metabolite present in the medium in relation to the total amount of product biosynthesized varies according to the type of molecule, volume of medium available, and also the cell strain: some strains excrete, and others do not. Within a collection of strains, the variability of this parameter may be exploited to select excreting strains.

Diffusion may be active. Tabata observed in electron microscopy that the naphthoquinones which constitute shikonin, synthesized at the endoplasmic reticulum of Lithospermum erythrorhizon cells, are secreted by exocytosis of pigmented granules.

A metabolite concentration gradient may be drawn up, such as the model proposed by Renaudin (1982) based on the study of the distribution balance of marked exogenous alkaloids. Distribution depends on the pK of the alkaloid in question, a pH gradient between vacuole, cytoplasma, cell wall and external medium, and differences in potential at the level of tonoplasma and plasmalemma.

Moreover, part of the alkaloids may form complexes with other substances, and this may be irreversible. This possibility of the desired metabolite complexing with other substances in the cell poses a dual problem: the correct evaluation of the biosynthesis capacities of the strain, and actual metabolite extraction. By forming complexes, these metabolites become inaccessible to classical extraction methods. Hutin (1983) showed on a particular strain of Papaver somniferum, that morphine and codeine did not appear in free form, as in the whole plant or in other tissue strains, but remained combined within the cell. Normal extraction as commonly used on the whole-plant poppy suggested the conclusion that this strain did not produce morphine alkaloids. But a particular type of extraction, in this case hot strong acid hydrolysis for one hour, allowed morphine and codeine to be isolated and identified. Great caution must therefore be exercised when evaluating the biosynthesis capacities of a given strain, before affirming that it is non-productive.

Let us sum up this chapter on the state of the technology with the answers to our three key questions:

It is now possible to select highly productive cell strains in order to obtain known and/or original compounds.

Cell strains may be cultivated in various types of bioreactor, by different processes, and on an industrial scale.

Certain metabolites may be extracted from the culture medium. In some cases, metabolites produced are accumulated within the cell in the form of unusual combinations requiring the use of special extraction techniques.

CURRENT ADVANCES - SOME EXAMPLES

The Development of New Molecules

The use of in vitro plant cell and tissue cultures as a research tool for new products is not subject to the same economic imperatives as when used as a production tool. At the beginning at least, little attention is paid to the expense of the process, the volume and extent of the market, in a word, the entire precise economic context allowing the competitiveness of one production pathway to be evaluated in function of the availability of the product by other channels. The search for new active principles requires access to the widest possible source of diversity, to increase the chances of discovering new substances. In this case, a highly diversified strain bank must be made available by exploiting the variability appearing in vitro, whether spontaneous, or induced by various means. Once this reservoir of potential new substances has been established, the right questions must be asked, i.e. biological testing allowing the identification of pharmacological activities. The success of such a research programme finally depends on the efficiency with which the new active principles can be fractionated, purified, and identified.

In this respect, the work of Arens (1982) is exemplary. The successive fractionation of extract of Picralima nitida cells cultivated in vitro allowed him to isolate two indole alkaloids with analgesic qualities: pericine and pericallin. The former of the two had until that time never been described. Starting with the seeds of this Apocynaceae, well known in the traditional Pharmacopoeia of the Ivory Coast for its various medicinal properties, he stabilized first of all in calli, then in suspension, various alkaloid-biosynthesizing strains. The pharmacological activity of the extracts was followed up by competition with naloxone in a binding test using opiate receptors taken from rat's brain, and both compounds isolated have a degree of activity on a par with that of codeine.

The work of Umetami et al. (1982) is an example of a different approach: the appearance of new molecules is favoured by giving the plant cell an unexpected substrate. About ten strains established from 7 different plant species were tested for their ability to glycosylate numerous products foreign to their metabolism. Salicylic acid, for example, was glycosylated with a high yield (61%) by a cell suspension of Mallotus japonicus. Mice were used to evaluate the analgesic activity of the new compound obtained. The activity was found to be on a par with that of salicylic acid, but much more precocious and long-lasting, probably due to the increased solubility of the product - obviously a useful property for a new medicine.

It should be borne in mind that, in certain cases, new molecules discovered in plant cell cultures might subsequently be produced by chemical synthesis. Production processes using bioconversion in plant cell cultures may be preferred for other reactions specific to plant cells and difficult by organic synthesis, e.g. glycosylation.

Shikonin Production

The first product obtained from mass plant cell culture was shikonin, a red pigment composed of eight naphthoquinones and extracted from the root of a Japanese plant called "Shikon" (Lithospermum erythrorhizon). This product has a biological activity resembling that of vitamin P, and is used in cosmetics and in pharmacy, as an externally applied anti-inflammatory for the treatment of burns. Ninety-five per cent of the raw material, the Shikon root, was imported from China, and the price of shikonin, 4500 $/kg, justified the attempt at production using biotechnology.

The project was set under way at the University of Kyoto (Tabata et al., 1974) roughly following the methodology outlined above. The first stage was the selection of highly productive cell lines, facilitated by the red colour of the product. This preliminary selection allowed a 25-fold increase in productivity. A factor of 2.8 was gained by the optimization of a culture medium favouring biosynthesis. Industrial scaling up and the definition of a specific growth medium multiplied productivity by a further 5.33 times, causing a 372-fold total gain in productivity from initial strain to industrial level. As growth and production media were not the same, a two-stage fermentation process was adopted. Production line productivity (750 l fermenting vat) is in the order of 0.4 kg/day (Fujita et al, 1983).

The present market, limited to Japan, represents approximately 15 kg/year, and the new selling price is 4,000 $/kg. The Japanese company Mitsui Petrochemicals hopes to extend the market to other countries and is preparing to double (or has already doubled) its production capacity. The "biolipstick" developed by Kanebo is, of course, only one of the possible applications of this medicinal pigment, but 1.3 million tubes at roughly 13 $ apiece, i.e. a turnover of 17 million dollars, had already been produced two months after this range was launched in February 1984, in part probably due to the "natural product" label linked by Kanebo with the beautiful red colour of this pigment. The same company has already added a shikonin-based cosmetic powder to its "Lady-Eighty" line, and is getting ready to produce a geranium perfume by plant cell culture.

One more word about the naphthoquinones shikonin is made up of: they are all present in cultivated cell extract, and the reproducibility of this extract is superior to that of extract taken from shikonin roots. Acetylshikonin content, for example, varies from 43 to 49% in extract obtained by cell culture, whereas for extract from roots, the variation goes from 34 to 63% depending on the origin and harvesting conditions of the raw material. The possibility of stable production is a strong argument in favour of biotechnology as opposed to other production channels.

Digoxin Production

The case of digoxin is a typical example of production processes using plant cell bioconversion.

Digitoxin and digoxin are very similar cardioactive substances, differing only by a hydroxyl group in position 12. They are extracted from Digitalis lanata and are used for various medicines, but pharmakinetics has since shown that digitoxin is in fact less useful than digoxin. In view of present production, a way was sought of changing digitoxin into digoxin, which has not proved possible by chemical means, nor by conversion using microorganisms.

Reinhard's group at Tübingen therefore sought to use plant cells for this conversion with the same basic methodology described above (Alfermann and Reinhard, 1978, 1980; Heins, 1978; Wahl, 1978). From a very wide selection of Digitalis strains, they isolated lines specifically capable of performing hydroxylation in C12 after eliminating many other strains unable to efficiently carry out the desired transformation or which gave other products. Among the strains performing the desired bioconversion, screening resulted in a 16-fold increase in digoxin yield. Productivity was doubled twice more by optimization of the medium, and culture in the bioreactor, to give a 62-fold global yield improvement compared to the initial yield, thus attaining a production rate of 47 mg/l per day.

Hydroxylation takes place entirely within the medium, and scaling up is initiated using a 35 l intermediate reactor for ten days, using highly reproducible samples under cryopreservation. Development is at present at the stage of the 200 l fermenter, in batch for 18 days or in semi-continuous with the regular addition of substrate. Eighty-two per cent of the digitoxin is converted into digoxin, with only just over 1% lost in unusable Purpurea glycoside A, underlining the high specificity of the selected strain. Approximately 5% of the substrate is catabolized, representing a considerable loss in view of its price (1,000 $/kg). Unused substrate, however, can be recovered and recycled, leaving a balance of 93.5% for glycosides.

Alfermann also developed another process, immobilizing _Digitalis_ cells in alginate beads (Moritz et al., 1982). These beads are introduced into a column reactor, in which cells continuously producing digoxin may survive for over 6 months. However, the productivity rate of about 10 mg/l per day is still 4 or 5 times lower than in a semi-continuous free cell culture.

Economically speaking, the increase in value thanks to this conversion reaction is in the order of 2,000 $/kg for a potential volume of 3.5 tonnes/year, which would correspond to an annual turnover of 7 million dollars. As far as we know, however, no digoxin produced in this way has as yet been put on the market.

Examples of Industrialization

While gathering information currently available on the industrial exploitation of plant cell culture (Table 6), we were aware that this information would be very quickly out of date. The situation is rapidly evolving. At the beginning of 1984, this table would only have contained a single product. It is equally remarkable that only two countries are represented, with the balance tilted heavily in the favour of Japan.

CONCLUSION

Thirty years will have been necessary, since the first patent application concerning the mass culture of plant cells was made by Nickell (1954, for Pfizer), to see these first steps become an industrial reality: thirty years during which time the development of this technique has been held back in France by a number of _a priori_ statements of the type "it's not possible to...", "we'll never be able to...", making industry wary of committing itself to onerous, high-risk research along these lines.

In other countries, for example West Germany and Japan, these prejudices had less of an impact, and this, linked perhaps with better interdisciplinary cooperation, allowed greater strides to be made in this domain.

The few examples shown here clearly underline the industrial reality of this new means of production of useful substances, and it is to be hoped that the incredulity which still reigned in industrial and scientific circles until two or three years ago with regard to the feasibility of such techniques has finally disappeared forever.

150

Table 6. Products Already Commercialized or Under Development

Product	Species	Application	Price (with reservations	Company	Country
Shikonin	Lithospermum erythrorhizon	Pharmacy Cosmetics Dyes	4,000 $/kg	Mitsui Petro- chemicals	Japan
Berberine	Coptis japonica Thalictrum minus	Pharmacy		Mitsui Petro- chemicals	Japan
Biomass	Panax ginseng	Dietetics	30 $/kg	Nitto Denki Kogyo	Japan
Peroxidase	Raphanus	Diagnosis	2,000 $/kg	Toyobo	Japan
Geraniol	Geranium	Perfumery		Kanebo	Japan
Rosmarinic acid	Coleus blumei	Pharmacy		Natterman	GFR
Digoxin	Digitalis lanata	Pharmacy	3,000 $/kg	Boehringer Mannheim	GFR

The question to be asked today is: which substances should be produced? The source of supply of plant-originating compounds whose value and size of market would justify this approach should be looked at anew, and the degree of competitiveness using cell cultures should be evaluated. The possible number of targets is still narrowly restricted: plant substances of real economic interest are few and far between.

The evolution of this technology will follow two complementary paths:

- improved techniques will cause costs to be reduced, thus making the method competitive for a greater number of substances;

- parallel to this, the technology in question will be self-justifying by allowing the discovery of new molecules since, as we have seen, the richness of higher plant species as a source of complex molecules is potentialized in in vitro culture by spontaneous or inducible variability of the biological material. The economic interest of this new source is very clear, especially in the domain of pharmaceutics.

The next few years will be crucial as far as the evolution of the industrial applications of this technology is concerned. In France, otherwise rival companies are collaborating on this front (the present work is an example of such collaboration). The aim should be to arouse a certain critical awareness vis-à-vis a domain to which France has made a major contribution at the fundamental research level, but has not as yet reaped any of the benefits.

REFERENCES

Abou-Mandour, A.A., Fischer, S. and Czygan, F.C., 1979, Regeneration von intakten pflanzen aus diploiden und haploiden kalluszellen von Catharanthus roseus, Z. Pflanzenphysiol., 91, 1:83-88.

Alfermann, A.W. and Reinhard, E., 1978, Possibilities and problems in production of natural compounds by cell culture methods, in: "Production of natural compounds by cell culture methods", A.W. Alfermann and E. Reinhard, eds., Munchen, pp 3-5.

Alfermann, A.W. and Reinhard, E., 1980, Biotransformation by plant tissue cultures, Bull. Soc. Chim. Fr., 2:35-45.

Arens, H., Borbe, H.O., Ubrich, B. and Stockigt, J., 1982, Detection of pericine, a new CNS active indole alkaloid from Picralima nitida cell suspension culture by opiate receptor binding studies, Planta Medica, 46:210-214.

Berlin, J., Beier, H., Fecker, L., Noe, W. and Schairer, H.U., 1984, Transformation of plant cell cultures for the increased formation of secondary metabolites, in: "Research and training programme in biomolecular engineering", progress report 1983, Vol. 1, Commission of the European Communities, Ed., ECSC - EEC - EAEC, Brussels, Luxembourg, pp 126-128.

Chang, W.C. and Hsing, Y.I., 1980, In vitro flowering of embryoids derived from mature root callus of Genseng (Panax ginseng), Nature, 284:341-342.

Dumanoir, J., Desmarest, P. and Saussay, R., 1985, In vitro propagation of fennel (Faeniculum vulgare Miller), Scientia horticulturae, submitted for publication.

Erdei, I., Kiss, Z. and Maliga, P., 1981, Rapid clonal multiplication of Digitalis lanata in tissue culture, Plant Cell Reports, 1:34-35.

Fujita, Y., Maeda, Y., Suga, C. and Morimoto, T., 1983, Production of shikonin derivatives by cell suspension cultures of Lithospermum erythrorhizon, Plant Cell Reports, 2:192-193.

Furuya, T., Yoshikawa, T. and Kiyohara, H., 1983, Alkaloid production in cultured cells of Dioscoreophyllum cumminsii, Phytochemistry, 22:1671-1674.

Guern, J., 1979, Les cellules de plante cultivées en milieu liquide et la croissance de leurs populations, in: "Production de substances naturelles par culture in vitro de tissus et de cellules de végétaux", Colloque A.P.R.I.A. - D.G.R.S.T., 21-22 mai 1979, Tours, pp 91-108.

Hahlbrock, K., 1974, "Correlation between nitrate uptake, growth, and changes in metabolic activities of cultured plant cells. Tissue, culture and plant science", Academic Press, London.

Hashimoto, T., Azechi, S., Sugita, S. and Suzuki, K., 1982, Large scale production of tobacco cells by continuous cultivation, in: "Plant tissue culture", A. Fujiwara, ed., Maruzen, Tokyo, pp 403-404.

Heins, M., 1978, Screening of Digitalis lanata plant and cell cultures for hydroxylation capacity, in: "Production of natural compounds by cell culture methods", A.W. Alfermann and E. Reinhard, eds., Munchen, pp 39-47.

Hiraoka, N. and Tabata, M., 1974, Alkaloid production by plants regenerated from cultured cells of Datura innoxia, Phytochemistry, 13:1671-1675.

Hutin, M., Foucher, J.P., Courtois, D. and Petiard, V., 1983, Evidences for unusual forms of storage of morphinan in a Papaver somniferum L., C.R. Acad. Sci. Paris, 297, III:47-50.

Misawa, M., Hayashi, M. and Tabata, S., 1983, Production of antineoplastic agents by plant tissue cultures. I - Induction of callus tissues and detection of the agents in cultured cells, Planta Medica, 49, 2:115-119.

Moritz, S., Alfermann, A.W. and Reinhard, E., 1982, Continuous biotransformation by immobilized cells in bioreactors, Planta Medica, 45:154-162.

Noguchi, M., Matsumoto, T., Hirata, V., Yamamoto, K., Katsuyama, A., Kato, A., Azechi, A. and Kato, K., 1977, Improvement of growth rates of plant cell cultures, in: "Plant tissue and its biotechnological application", W. Barz, E. Reinhard and M.H. Zenk, eds., Springer-Verlag, Berlin, pp 85-94.

Petiard, V., Courtois, D., Gueritte, F., Langlois, N. and Mompon, B., 1982, New alkaloids in plant tissue culture, in: "Plant tissue culture", A. Fujiware, ed., Maruzen, Tokyo, pp 309-310.

Petiard, V. and Courtois, D., 1983, Recent advances in reserach for novel alkaloids in Apocynaceae tissue cultures, Physiol. Veg., 21:217-227.

Petiard, V. and Courtois, D., 1984, unpublished results.

Renaudin, J.P. and Guern, J., 1982, Compartmentation mechanisms of indole alkaloids in cell suspension, Physiol. Veg., 20:533-547.

Seitz, U., 1985, Storage of plant tissue cultures, Société Française de Microbiologie - Xe Colloque annuel. Les aspects industriels des cultures cellulaires d'origine animale et végétale. INSA-LYON, 7 et 9 mars 1985, to be published.

Shimomura, K., Teshima, D., Soyama, Y. and Nishioka, I., 1980, Shoyakuga Zasshi, 34:306.

Shono, K. and Furuya T., 1972, Experimentia, 28:236.

Shoyama, Y., Hatano, K. and Nishioka, I., 1983, Clonal multiplication of Pinellia ternata by tissue culture, Planta Medica, 49:14-16.

Staba, E.J., 1969, Plant tissue culture as a technique for the phytochemist, Recent Adv. Phytochemistry, 2:75-106.

Stockigt, J., Pfitzner, A. and Firl, J., 1981, Indole alkaloids from cell suspension cultures of Rauwolfia serpentina Benth, Plant Cell Reports, 1:36-39.

Tabata, M., Mizukami, H., Hiraoka, N. and Konoshima, M., 1974, Pigment formation in callus cultures in Lithospermum erythrorhizon, Phytochemistry, 13:927-932.

Tulecke, W. and Nickell, L.G., 1960, Methods, problems and results of growing plant cells under submerged conditions, Tans. New York Acad. Sci., 22:196-206.

Umetami, Y., Tanaka, S. and Tabata, M., 1982, Gl cosylation of extrinsic compounds by various plant cell cultures, in: "Plant tissue culture", A. Fujiwara, ed., Maruzen, Tokyo, pp 383-384.

Wahl, J., 1978, Biotransformation of B-methyl-digitoxin by Digitalis lanata fermenter cultures, in: "Production of natural compounds by cell culture methods, A.W. Alfermann and E. Reinhard, eds., Munchen, pp 48-49.

Waller, G., MacVean, C. and Suzuki, T., 1983, High production of caffeine and related enzyme activities in callus cultures of coffea arabica L., Plant Cell Reports, 2:109-112.

Yamada, Y., 1984, in: "Primary and secondary metabolism of plant cell cultures", Springer-Verlag, Giessen, to be published.

Yasuda, S., Satoh, K., Ishii, T. and Furuya, T., 1972, Studies on cultural conditions of plant cell suspension cultures, Proceedings of 4th Int. Symp. of Fermentation, Kyoto.

Zenk, M.H., El-Shagi, H., Arens, H., Stockigt, J., Weiler, E.W. and Deus, B., 1977, Formation of the indole alkaloids serpentine and ajmalicine in cell suspension cultures of Catharanthus roseus, in "Plant tissue culture and its biotechnological application", W. Barz, E. Reinhard and M.H. Zenk, eds., Springer-Verlag, Berlin, pp 27-44.

Zenk, M.H., 1982, Large scale culture and productivity of plant cells, Plant Cell Culture Conference, London.

CLONING OF GENES IN STREPTOMYCETES FOR SECONDARY METABOLISM

Teruhiko Beppu and Sueharu Horinouchi

Department of Agricultural Chemistry
The University of Tokyo, Bunkyo-ku, Tokyo 113, Japan

SUMMARY

Streptomycetes are Gram-positive bacteria with an extremely high guanine plus cystosine content (70 to 73%). Their capability to produce the great majority of secondary metabolites including most of the antibiotics makes them industrially important. A complex process of morphological differentiation displayed by Streptomyces also has biologically interesting aspects. A close relationship of secondary metabolism with cell differentiation is well recognized in streptomycetes. It seems reasonable to assume that multiple genes involved in both the complex processes are controlled by a common regulatory gene or substance.

A representative of such regulatory substances is A-factor (2-isocapryloyl-3R-hydroxymethyl-γ-butyrolactone; Mori, 1983) which was originally found by Khokhlov et al. (1967) in the culture broth of Streptomyces griseus (Figure 1). A-factor is a self-regulatory substance or bioregulator essential for streptomycin production, streptomycin resistance, and spore formation in this organism (Khokhlov et al., 1973; Khoklov, 1980; Hara and Beppu, 1982a, 1982b). A-factor-deficient mutants of S. griseus simultaneously lose streptomycin production and resistance, and spore forming ability. In such mutants, addition of A-factor at a concentration of 10^{-9} M restores all the defects. These features of A-factor are similar to hormones in eukaryotes.

Recent development of host-vector systems for streptomycetes has enabled us to clone various genes involved in secondary metabolism. In order to clarify the genetic background of the A-factor regulatory system, we have cloned and characterized an A-factor determinant as well as a streptomycin synthesizing gene as a possible target of the A-factor function

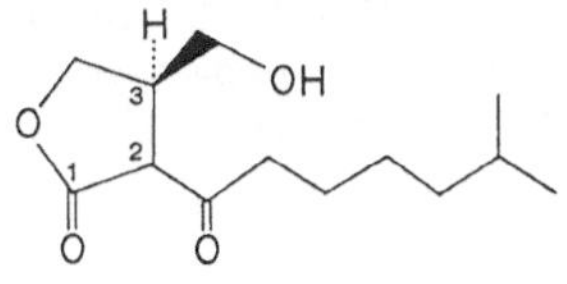

Fig. 1. Chemical structure of A-factor

from streptomycin-producing <u>Streptomyces bikiniensis</u>. In the course of the experiments, we also found that a regulatory gene, <u>afsB</u>, positively controlling biosynthesis of A-factor in <u>Streptomyces coelicolor</u> A3(2) simultaneously regulated biosynthesis of pigmented antibiotics actinorhodin and undecylprodigiosin in this organism as well as in <u>Streptomyces lividans</u>. We will describe below cloning and characterization of these genes involved in secondary metabolism.

STREPTOMYCES BIKINIENSIS afsA GENE

Streptomycin production and spore formation of <u>S. griseus</u> and its relative, <u>S. bikiniensis</u>, are easily lost by acridine orange treatment and by incubation at high temperature (Hara and Beppu, 1982a). We found that these defects were in fact due to loss of A-factor production. This observation suggested the presence of an unstable and extrachromosomal genetic determinant of A-factor biosynthesis. Our genetic mapping of A-factor genes in <u>S. griseus</u> by using protoplast fusion techniques revealed an extrachromosomal nature, in which A-factor genes do not show linkages with any marker on the chromosome and furthermore almost all of the fusants showed an A-factor-producing phenotype (Hara et al., 1983). Therefore, we have assumed that the A-factor determinant in this organism and also in <u>S.-bikiniensis</u> is carried on an extrachromosomal element, possibly a plasmid.

Cloning of S. bikiniensis A-factor determinant

In order to clone the A-factor determinant in streptomycin-producing organisms, we used an A-factor-deficient mutant strain <u>S. bikiniensis</u> HH1 obtained by incubation at 37°C as the host, which was transformable at higher frequency than <u>S. griseus</u>, and neomycin and thiostrepton resistance plasmid pIJ385 as the cloning vector. As donor DNA, we used total cellular DNA from <u>S. bikiniensis</u> to avoid possible modification-restriction barriers (Horinouchi et al., 1984). Selection of transformants carrying the A-factor determinant was based on the pleiotropic effect of A-factor on streptomycin production; restoration of A-factor production subsequently triggers streptomycin production in the host. We detected one colony producing streptomcyin among more than 20,000 thiostrepton-resistant transformants. The isolate contained a recombinant plasmid, named pAFB1, with a 9.0-kilobase pairs (kb) insertion (Figure 2).

Phenotypic Expression of the Cloned Fragment

Purified pAFB1 DNA was reintroduced by transformation into <u>S. bikiniensis</u> HH1 to test for phenotypic expression in the host cell. All the thiostrepton-resistant transformants obtained in this way were confirmed to produce streptomycin as well as to form spores. In a bioautogram of chloroform extracts of culture filtrates of <u>S. bikiniensis</u> (pAFB1), a strongly positive spot with an <u>Rf</u> value identical to that of chemically synthesized A-factor was detected. The amount of A-factor produced by the plasmid-carrying HH1 was 18 times larger than that produced by the parental strain, probably due to gene dosage effect of the multicopy vector plasmid pIJ385 (40 to 300 copies per chromosome). Plasmid pAFB1 also conferred A-factor production in a large quantity to A-factor-deficient mutants of <u>S. griseus</u> strain FT-1 and strain IFO 13189 obtained by "plasmid curing treatments", as a result of which streptomycin production, streptomycin resistance, and spore formation of these mutants were simultaneously recovered.

Biosynthesis of A-factor in <u>S. coelicolor</u> A3(2) is determined by two chromosomal loci, <u>afsA</u> and <u>afsB</u> (Hara et al., 1983), as described below. Mutations at <u>afsA</u> cause only loss of A-factor production, whereas <u>afsB</u>

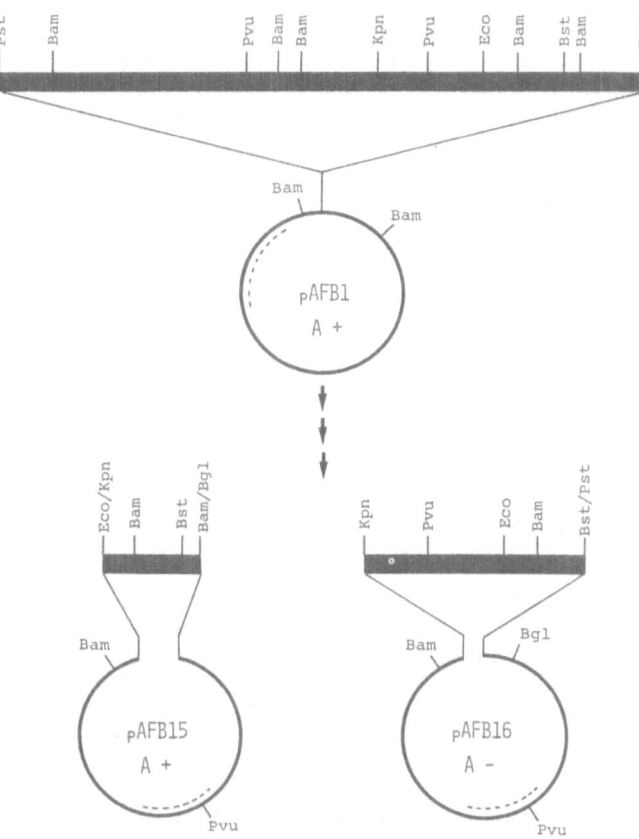

Fig.2. Structures of recombinant plasmids. Originally isolated plasmid pAFB1 contains a 9.0-
kb DNA fragment in the Pst1 site 05 the
vector plasmid pIJ385. Plasmid pAFB15
contains a trimmed 1.1-kb fragment with the
ability to confer A-factor production.
Plasmid pAFB16 lacking a short BstEII-BamHI
fragment failed to confer A-factor production

mutants lose biosynthetic activities not only A-factor but also of the
pigments actinrhodin and undecylprodigiosin. We presume that afsA is a
structural gene(s) for A-factor biosynthetic enzyme(s) and afsB, a
regulatory gene involved in secondary metabolism. Plasmid pAFB1 also
conferred A-factor production in a large quantity to both afsA and afsB
mutants of S. coelicolor A3(2). On the other hand, production of the
pigments was not restored in the afsB mutants.

Characterization of the A-factor Determinant by Southern Blot Analysis

Genetic analysis in S. griseus showed nonlinkage of the A-factor
determinant to chromosomal markers. In addition, its infectious transfer
upon protoplast fusion suggested involvement of an extrachromosomal genetic
element of A-factor biosynthesis in this organism. To determine the
presence or absence of A-factor determinants in A-factor-deficient mutants
of S. griseus and S. bikiniensis, we performed Southern blot experiments
with the ^{32}P-labeled fragment containing the intact A-factor gene(s) as
probe and total cellular DNA from A-factor-negative mutants, as well as

their parental strain, as targets (Horinouchi et al., 1984). The parental
S. griseus strains had positive hybridization bands with the same intensity
and the same BamH1 restriction pattern as those of S. bikiniensis. These
results suggested that the nucleotide sequences of the A-factor gene(s) of
the two strains were well conserved. However, A-factor-deficient mutants of
S. bikiniensis and S. griseus gave no positive hybridization. These results
indicate that the entire DNA sequences homologous to the cloned A-factor
determinant of S. bikiniensis are easily lost as a unit in these organisms,
in accordance with the assumption that the A-factor genes are carried on an
unstable extrachromosomal element in streptomycin-producing streptomycetes.

A-factor production has been found to be distributed widely among
actinomycetes (Khokhlov, 1980; Hara and Beppu, 1982a). To analyze
distribution of the A-factor gene(s) in actinomycetes, we hybridized the
above probe with BamHI-digested total cellular DNA extracted from various
other A-factor-producing and nonproducing strains. Most of the A-factor
producers carried a sequence homologous to the probe, with a varying degree
of homology with different BamHI restriction patterns, whereas most of the
nonproducers did not (Figure 3). These findings suggest that the A-factor
determinant of S. bikiniensis and those of other actinomycetes are related
and that they have diverged from a common ancestral sequence.

Trimming the A-factor Determinant by Subcloning

It is of interest to reveal how many genes are required for A-factor
biosynthesis. For that, we trimmed the cloned 9.0-kb fragment by subcloning
and identified a transcriptional control signal on the trimmed fragment
(Horinouchi et al., 1985).

Figure 2 shows the schematic diagram for subcloning the A-factor
determinant by using pAFB1 as the starting material. Plasmid pAFB15
containing a trimmed 1.1-kb fragment was capable of conferring A-factor

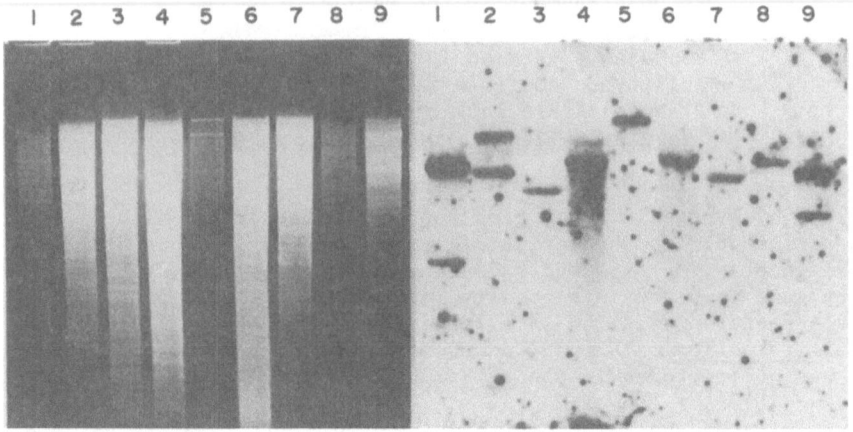

Fig. 3. Agarose gel showing total DNAs digested with BamHI
left panel) and autoradiogram of DNA/DNA hybridi-
zations between BamHI-digested total DNAs of A-
factor-producing strains and the cloned S. bikiniensis
A-factor determinant (right panel). 1. S. albus IFO 3195;
2. S. antibioticus ID.FO 3126; 3. S. antibioticus IFO
12652; 4. S. antibioticus IFO 12838; 5. S. flaveolus
IAM 0117; 6. S. globisporum IFO 12208; 7. S. griseo-
flavus IFO 12372; 8. S. sindensis IFO 12915;
9. A. citreofluorescens IFO 12853

production to <u>S. bikiniensis</u> HH1. The amount of A-factor produced by pAFB15-carrying strain HH1 was as large as that of pAFB1-carrying HH1 (3.0μg/ml), in comparison with the parental strain which produced about 0.3μg/ml of A-factor. Plasmid pAFB15 also conferred A-factor production in a large quantity to A-factor-deficient mutants of <u>S. griseus</u> as well as <u>afsA</u> mutants of <u>S. coelicolor</u> A3(2). Furthermore, four <u>Streptomyces</u> strains which orginally had no ability to produce A-factor and no DNA sequence homologous to the <u>S. bikiniensis</u> A-factor determinant began to produce A-factor with a remarkable gene dosage effect upon introduction of pAFB15.

Presence of a long open reading frame deduced from nucleotide sequencing of the 1.1-kb fragment suggested the orientation of the A-factor determinant was from the <u>EcoRI</u> site to the <u>BamHI</u> site (unpublished results). A promoter-probe plasmid pARC1 with a unique <u>BamHI</u> cloning site contains a chromogenic gene(s) on the vector plasmid pIJ41 and allows identification of transcriptional signals by pigment production (Horinouchi and Beppu, 1985). When a DNA fragment with transcriptional control signals is inserted in the correct orientation in the <u>BamHI</u> site of plasmid pARC1, production of a brown pigment, presumably a shunt product in the actinorhodin biosynthetic pathway is induced. The 450-bp <u>EcoRI</u>-<u>BamHI</u> fragment from the cloned A-factor determinant was inserted into pARC1 digested with <u>HindIII</u> plus <u>BamHI</u>, resulting in pARC4, as shown in Figure 6. Deletion of the <u>BamHI</u> to <u>HindIII</u> fragment of pARC1 does not increase read-through transcription from the vector sequence, if any, to such an extent that pigment production can be detected. Transformants carrying pARC4 produced a brown pigment in a large quantity, which revealed the presence of a promoter signal whose direction is from the <u>EcoRI</u> site to the <u>BamHI</u> site. In another experiment in which the 650-bp <u>BamHI</u>-<u>BamHI</u> (the remainder) fragment in the 1.1-kb fragment was inserted into the <u>BamHI</u> site of pARC1 in two orientations, none of the transformants showed a pigment-producing phenotype.

From all the results described above, it is evident that the 1.1-kb fragment contains both promoter and coding sequences capable of A-factor biosynthesis in a wide variety of streptomyces. From its small size, it appears likely that the fragment contains only one gene. It is most likely that the putative gene is a structural gene encoding for an enzyme involved in A-factor biosynthesis. Complementation of the <u>afsA</u> function of S. coelicolor A3(2) by the cloned fragment and a marked overproduction of A-factor by introducing the fragment on the multicopy vector plasmid in all the <u>Streptomyces</u> strains so far tested also support this assumption. These data also suggest that precursors for A-factor biosynthesis are common metabolites in streptomyces and that acquisition of only a single key enzyme encoded by the trimmed 1.1-kb fragment is sufficient for <u>Streptomyces</u> strains to synthesize A-factor. Nucleotide sequence determination of this region is now in progress.

CLONING OF STREPTOMYCIN SYNTHESIZING GENE(S) OF S. BIKINIENSIS

As a target gene of the regulatory functions of A-factor, we cloned streptomycin synthesizing gene(s) from <u>S. bikiniensis</u> by using a streptomycin-nonproducing <u>S. bikiniensis</u> probably having a defect in N-methyl-L-glucosamine biosynthesis (one of the moieties of streptomycin) as the host (unpublished results). The parent strain of <u>S. bikiniensis</u> accumulated both streptomycin and streptidine-dihydrostreptose in the media, but the mutant accumulated only the latter. A DNA fragment of 9.5-kb size complementing the defect of the mutant strain, was cloned using the multicopy vector plasmid pIJ385 (40 to 300 copies per chromosome). The host strain harbouring the hybrid plasmid recovered streptomcyin production with 7-fold enhancement in comparison with the parental strain. We are currently

examining the regulatory mechanism of A-factor on the cloned streptomycin
production gene(s).

A PLEIOTROPIC REGULATORY GENE, afsB, OF S. COELICOLOR A(3)2 INVOLVED
IN SECONDARY METABOLISM

    As mentioned above, A-factor production is widely distributed not only
in streptomycin-producing organisms but also in various actinomycetes
including S. coelicolor A3(2), genetically the most intensively studied
strain. By using the well developed gene exchange systems of this organism
through conjugation, A-factor nonproducing mutations were mapped on the
chromosome between cycD and leuB (Hara et al., 1983), in contrast to the A-
factor determinant of streptomycin-producing organisms which had been
suggested to be extrachromosomal. These analyses also suggested two
slightly different chromosomal positions for two types of afs (termed for A-
factor synthesis) mutations. We termed afsA for one of the mutations which
lacked A-factor production only, and afsB for the other which was
characterized by concomitant loss of A-factor and of pigments actinorhodin
and undecylprodigiosin. The involvement of these two genes, afsA and afsB,
for A-factor biosynthesis in S. coelicolor A3(2) and S. lividans was
confirmed by cloning experiments described below (Horinouchi et al., 1983).

    The cloned afsB gene which was narrowed down to about 2-kb in length
restored A-factor and pigment deficiencies of S. coelicolor A3(2) afsB
mutants. Furthermore, introduction of the AfsB gene into S. lividans (both
A-factor-deficient strain HH21 and a wild type strain TK21) caused,
unexpectedly, production of pigments actinorhodin and undecyl prodigiosin in
a large quantity (Horinouchi and Beppu, 1984). We describe the pleiotropic
regulatory features and nucleotide sequence of the cloned afsB.

Cloning of the S. coelicolor A3(2) afsB Gene

    To examine the apparent discrepancy between the genetic behaviours of
the A-factor genes in S. coelicolor A3(2) and streptomycin-producing
organisms, we tried to clone A-factor genes of S. coelicolor A3(2) by using
the host-vector system of S. lividans. A plasmid vector pIJ41 was used as
the vector and a spontaneously derived A-factor-deficient strain (HH21) of
S. lividans was used as the host for the shotgun cloning of the afs genes of
S. coelicolor A3(2). Detection of A-factor-positive transformants failed
because of appearance of the revertants producing A-factor. Therefore, we
applied an experimental observation with S. coelicolor A3(2) strain that A-
factor production might be associated with red pigment production, and
selected red-colored S. lividans transformants from the gene bank of the
S. coelicolor A3(2) chromosomal DNA. Three transformants obtained in this
way were found to produce A-factor and carry a recombinant plasmid with a
6.4-megadalton (about 9-kilobase) insertion. By subcloning, the region
necessary for endowing A-factor production to S. lividans HH21 was narrowed
down to about 2-kb fragment. Introduction of the 2-kb fragment into afs
mutants of S. coelicolor A3(2), where the fragment complemented only the
afsB mutations but not afsA mutations, indicated that the cloned gene
corresponded to the afsB.

Expression of the Cloned Gene in S. lividans

    Plasmid pIJ41-AP3 containing 4.5-kb fragment covering the trimmed 2-kb
fragment is one of the recombinant plasmids carrying the afsB gene which
induces the production of both A-factor and red pigments in S. lividans.
Stimulation of pigment production by the afsB was proved not to be due to A-
factor, since addition of A-factor at various concentrations to S. lividans

160

did not cause pigment production. Spectral and chromatographic analyses of the red pigments showed that they were actinorhodin and undecylprodigiosin, both of which are pigmented antibiotics produced by S. coelicolor A3(2). These pigments were not produced by the original host on either complete or Bennett agar medium. Time course of actinorhodin production by S. lividans (pIJ41-AP3) was shown in Figure 4. The amount of undecylprodigiosin produced by the plasmid-carrying S. lividans was calculated to be about 170µg per gram of mycelia grown for 2 weeks on Bennett agar medium, while the plasmid-free strain did not produce a detectable amount of undecylprodigiosin.

The cloned S. coelicolor A3(2) afsB regulates in a positive manner the biosynthesis of A-factor, actinorhodin, and undecylprodigiosin. In an afsB mutant of S. coelicolor A3(2), productions of a calcium-dependent antibiotic (Lakey et al., 1983) and methylenomycin (Kirkby and Hopwood, 1977), in addition to the above three metabolites, appeared to be signficantly reduced (D.A. Hopwood, personal communication). These observations suggest a general stimulatory function of afsB for the secondary metabolism. The biosynthetic genes for these metabolites, except for methylenomycin, are dispersedly located on the linkage map of the chromosome (Rudd and Hopwood, 1979, 1980; Hara et al., 1983; Hopwood and Wright, 1983). The whole pathway for methylenomycin biosynthesis is located on a fertility plasmid (Wright and Hopwood, 1976; Aguilar and Hopwood, 1982). In addition, the chemical structures and biosynthetic pathways of these substances are totally different. Therefore, it is reasonable to assume that a cytoplasmic regulator, possibly a protein, encoded by the cloned afsB gene plays a regulatory role for the biosynthesis of these secondary metabolites.

Nucleotide Sequence of the S. coelicolor A3(2) afsB

The nucleotide sequence of the trimmed 2-kb afsB gene was determined by the Maxam-Gelbert method (Horinouchi et al., 1985), in which coding sequence for the putative AfsB protein consisting of 277 amino acids was found.

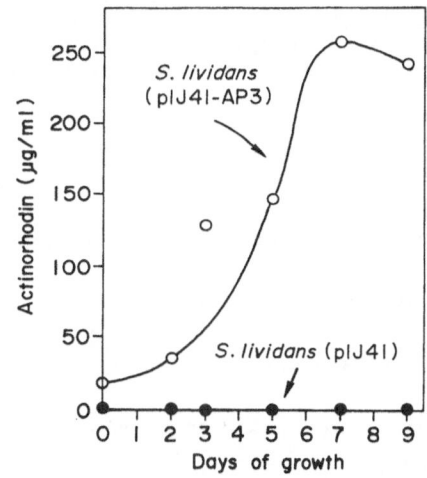

Fig. 4. Time course of actinorhodin production by S. lividans carrying pIJ41-AP3. The calculated amounts of actinorhodin in the culture broth was plotted together with those for S. lividans (pIJ41) as a reference.

161

Figure 5 shows the 5' region of the <u>afsB</u> gene. Based on the fine restriction map deduced from the nucleotide sequence, we subcloned various restriction fragments and found that a long reading frame (277 amino acid) starting with methionine (AUG) codon was responsible for <u>afsB+</u> phenotype. The putative AfsB protein deduced from the nucleotide sequence is shown in Figure 5. The codon usage is summarized in Table 1. An overall average guanine plus cytosine content of the coding sequence is 73 mol% and those for codon positions 1, 2, and 3 are 70.8, 56.7, and 92.1 mol%, respectively. The distribution of GC pairs at each of the three positions within the codons is in agreement with a feature of most <u>Streptomyces</u> genes as described by Binn et al. (1983). Table 2 summarizes the results, together with the data for the neomycin phosphotransferase (<u>aph</u>) and viomycin phosphotransferase (<u>vph</u>) genes for comparison. The high G+C composition is reflected preferentially in the third codon position as well as in the first position.

It is known that a degree of complementarity between the 3' end of the 16S rRNA of the prokaryotes and a nucleotide sequence situated several nucleotides upstream from the translational initiation codon of mRNA (the Shine-Dalgarno sequence) is necessary for the initiation of translation (Shine and Dalgarno, 1974; Gold et al., 1981). Comparing this with the nucleotide sequence of the 3' end of the 16S rRNA of <u>S. lividans</u>, 5'GAUCACCUCCUUUCU$_{OH}$ 3' (Bibb and Cohen, 1982), we infer that GAGG (nucleotides 594-597) or AGG (nucleotides 606-608), located 14 nucleotides or 4 nucleotides upstream in front of the presumptive ATG start codon, respectively, can serve as the Shine-Dalgarno sequence for the putative AfsB protein.

We also determined the transcriptional start site of <u>afsB</u> gene by SI nuclease mapping. The start site was 77 bp upstream from the putative <u>AfsB</u> translational initiation codon. The nucleotide sequence upstream of the mRNA start site did not contain "consensus" sequences for promoters found in other prokaryotes (Rosenberg and Court, 1979). This is also true of the <u>aph</u> gene (Thompson and Gray, 1983). Westpheling et al., (1985) reported <u>in vitro</u> transcription experiments with purified <u>S. coelicolor</u> A3(2) RNA polymerases using the <u>Streptomyces plicatus</u> endoglycosidase H (<u>endo H</u>) gene as well as the <u>B. subtilis ctc</u> promoter as templates. According to their results, one ($E\alpha^{41}$) out of at least two types of RNA polymerases ($E\alpha^{35}$ and $E\alpha^{41}$)

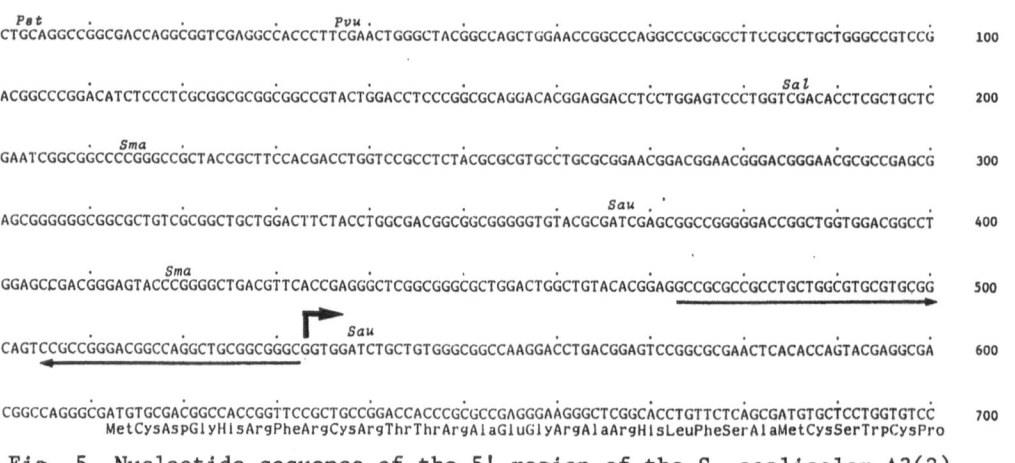

Fig. 5. Nucleotide sequence of the 5' region of the <u>S. coelicolor</u> A3(2) <u>afsB</u> gene. The thick arrow indicates the mRNA start site determined by S1 nuclease mapping.

Table 1. Codon usage of the putative AfsB protein

Arg (36)		Leu (28)		Ser (16)	
CGT	1	TTG	2	TCT	0
CGC	12	TTA	0	TCC	5
CGG	14	CTT	0	TCG	6
CGA	2	CTC	10	TCA	1
AGG	7	CTG	16	AGT	0
AGA	0	CTA	0	AGC	4

Thr (8)		Pro (7)		Val (13)	
ACT	0	CCT	0	GTT	0
ACC	5	CCC	3	GTC	8
ACG	3	CCG	4	GTG	4
ACA	0	CCA	0	GTA	1

Ala (45)		Gly (30)		Ile (3)	
GCT	3	GGT	1	ATT	0
GCC	23	GGC	21	ATC	3
GCG	15	GGG	2	ATA	0
GCA	4	GGA	6		

Asn (9)		Gln (11)		Tyr (2)	
AAT	0	CAG	11	TAT	0
AAC	9	CAA	0	TAC	2

His (11)		Glu (13)		Cys (9)	
CAT	0	GAG	13	TGT	3
CAC	11	GAA	0	TGC	6

Asp (11)		Phe (8)		Lys (4)	
GAT	0	TTT	0	AAG	4
GAC	11	TTC	8	AAA	0

Met (7)		Trp (6)	
ATG	7	TGG	6

Table 2. Distribution of nucleotides in the putative afsB, aph, and vph DNA sequences

Gene	Sequences (length)	G + C (mol %) in coden position[a]			
		1	2	3	Mean
afsB	Precoding (609 bp)	77	62	87	75
	Coding (831 bp)	71	56	92	73
	Postcoding (405 bp)	67	59	74	67
aph [b]	Precoding (306 bp)	85	67	84	79
	Coding (804 bp)	78	43	97	73
	Postcoding (170 bp)	80	82	84	82
vph [b]	Precoding (96 bp)	53	66	66	62
	Coding (861 bp)	79	51	91	74
	Postcoding (162 bp)	78	78	72	76

a The codon register is in frame with the presumptive ATG translational-start codons of the afsB, aph, and vph genes.

b The data for the aph and vph are taken from Bibb et al. (1983).

transcribed both the <u>endo</u> <u>H</u> and <u>ctc</u> promoters. The nucleotide sequences of these regions are shown in Figure 6, together with the corresponding region of <u>afsB</u>. Alignment of the <u>afsB</u> sequence revealed homology with the "-35" and "-10" sequences of the <u>ctc</u> promoter. The corresponding <u>afsB</u> sequences were significantly homologous to the <u>endo</u> <u>H</u> promoter. Computer analysis of the nucleotide sequence revealed the presence of an inverted complementary repeat sequence spanning a 60 nucleotide sequence at nucleotides 473 to 532 (ΔG=-57.2 kcal/mol). This stem-loop structure wholly contains the presumptive -35 and -10 regions. The role of the secondary structure is not clear.

Determination of afsB Promoter Region

To determine the region essential for expression of <u>afsB</u> function, we constructed various hybrid plasmids containing the presumptive AfsB coding sequence with upstream regions of different lengths, as shown in Figure 7. All the plasmids contained the AfsB coding sequence in the opposite orientation from that of the <u>aph</u> promoter on the vector to avoid a read-through transcription into the AfsB coding region. The structures of all the constructed plasmids were confirmed by purifying the plasmid DNAs followed by analyses of the cleavage patterns with appropriate enzymes. Plasmid pIJ41-AP47 containing only the <u>Sau</u>3A fragment with the whole coding sequence and plasmid pIJ41-ΔSau lacking the mRNA start site in the upstream region failed to confer A-factor production to <u>S. lividans</u> HH21. Neither pIJ41-AP43 nor pIJ41-AP45 containing 318-bp and 349-bp fragments upstream from the mRNA start site, respectively, conferred A-factor production. On the other hand, plasmids pIJ41-AP48 and pIJ41-AP49 containing 497-bp and 532-bp upstream from the 5' end of mRNA, respectively, conferred A-factor production to strain HH21. These results suggested that the <u>afsB</u> required a considerably long upstream sequence for the expression in <u>S. lividans</u>.

Then we constructed a similar series of plasmids by using plasmid pIJ702 as the vector to examine whether a similar conclusion was obtained in an <u>afsB</u> mutant strain BH6 of <u>S. coelicolor</u> A3(2) which lost the ability to produce A-factor as well as pigments actinorhodin and undecylprodigiosin owing to the deficiency of <u>afsB</u>. These experiments with <u>S. coelicolor</u> A3(2) also suggested that a considerably long sequence (about 500-bp) upstream from the mRNA start site is necessary for expression of the <u>afsB</u>, as the case with <u>S. lividans</u>.

Distribution of afsB Sequence among Streptomyces

Sequences homologous to the <u>afsB</u> gene with a varying extent of homology were found in several <u>Streptomyces</u> strains by Southern blot DNA/DNA hybridization in which the 2-kb <u>afsB</u> sequence was used as ^{32}P-labeled probe. It is not clear whether they represent a regulatory gene for secondary metabolism in those strains. Among 3 streptomycin-producing strains tested,

```
ctc     T T T C G A G G T T T A A A T C C T T A T C G T T A T G G G T A T T G T T T G T A A T A G
endo H  A T T G A C T G A T T G A C G C G C     T T C C G G C G G G C A G G G G A G G C A C G G T G
afsB    C G T G C G T G C G G C A G T C C G     C C G G G A C G G C C A G G C T G C G G C G G G C G
                              *   *             *     *               *               *
```

Fig. 6. Alignment of promoter sequences of the <u>ctc</u>, <u>endo</u> <u>H</u>, and <u>afsB</u> gene.
Underlined nucleotides in the <u>ctc</u> promoter are important for
utilization by <u>B. subtilis</u> RNA polymerase (Eα^{37}), shown by methy-
lation protection experiments (Moran et al., 1982). The
transcription start sites are shown by double lines. Identical
nucleotides of the <u>ctc</u> and <u>afsB</u> are asterisked.

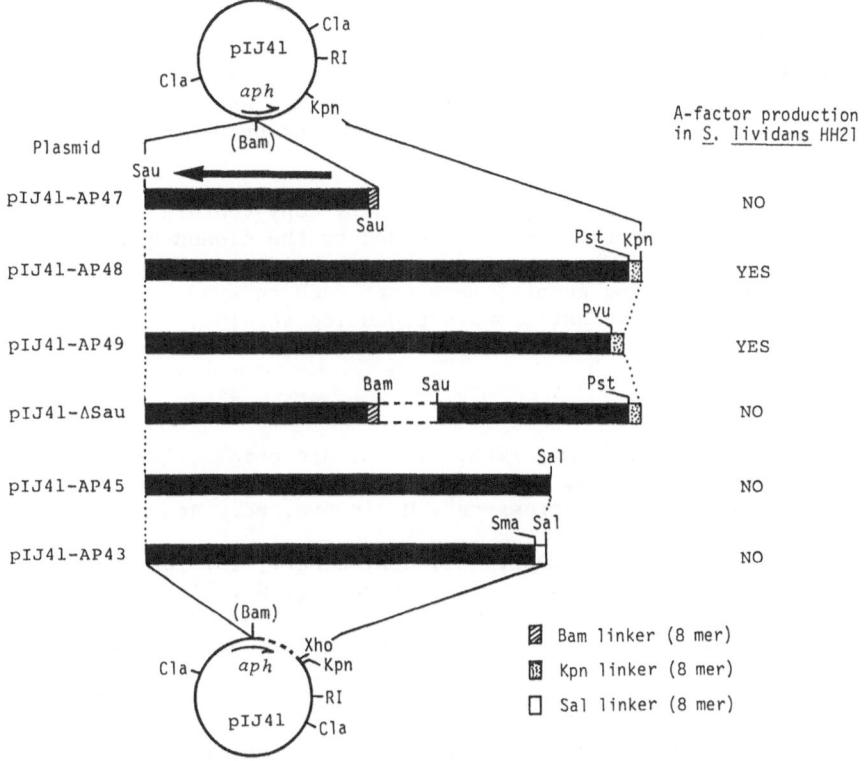

Fig. 7. Schematic representation of pIJ41 series plasmids and
their phenotypes determined by test for A-factor
production in S. lividans HH21. The thick arrow on
the map of pIJ41-AP47 stands for the putative AfsB
protein

only S. griseus FT-1 which is a high streptomycin-producer contained a
sequence considerably homologous to the afsB gene. On the other hand, A-
factor-deficient mutants derived by so-called plasmid curing treatments from
FT-1 were found to lack the homologous sequence. These results strongly
suggest that the entire sequence homologous to the afsB in this strain is
easily lost by curing treatments. Such behaviour suggests that the afsA
gene possibly encodes an A-factor biosynthetic enzyme. We infer that the
homologous sequence to the afsB is carried on the same plasmid carrying the
afsA or on a different plasmid. Possible regulatory roles of the homologous
sequence in S. griseus FT-1 should be clarified by further experiments.

CONCLUSION

Antibiotic production and cell differentiation in Streptomyces are
regulated by a complex mechanism in which two different levels of regulation
are functioning; one at the gene level, like afsB, and the other at the
level of a regulatory substance, like A-factor. Such a pleiotropic
regulatory network plays an important role for complex phenotypic expression
in streptomyces. The unexpected pigment production induced by the introduction of afsB gene suggests that the gene awakens "sleeping" genes of
S. lividans. This seems to indicate an important genetic feature of
streptomycetes, i.e., a large number of genes responsible for secondary
metabolism might be intrinsically present but most of them might be in the

dormant state in a <u>Streptomyces</u> strain. Such a feature will explain why the streptomycetes produce diverse secondary metabolites which differ from strain to strain. Cloning and characterization of such genes may be useful for achieving enhancement of antibiotic production and sometimes for finding novel antibiotics in various <u>Streptomyces</u> strains by awakening "sleeping" genes.

Gene amplification by cloning with multi-copy vectors frequently causes increased production of the protein encoded by the cloned gene, which in turn causes improved production of metabolites. Our results with a streptomycin-synthesizing gene(s) show that such an approach is also possible for breeding new antibiotics-producing strains.

REFERENCES

Aguilar, A. and Hopwood, D.A., 1982, J. Gen. Microbiol., 128:1893-1901.
Bibb, M.J., Chater, K.F. and Hopwood, D.A., 1983, in: "Experimental manipulation of gene expression", M. Inouye, ed., Academic Press, New York, pp 53-82.
Bibb, M.J. and Cohen, S.N., 1982, Mol. Gen. Genet., 187:265-277.
Gold, L., Pribnow, D., Schneider, T., Shinedling, S., Singer, B.S. and Stormo, G., 1981, Annu. Rev. Microbiol., 35:365-403.
Hara, O. and Beppu, T., 1982a, J. Antibiot., 35:349-358.
Hara, O. and Beppu, T., 1982b, J. Antibiot., 35:1208-1215.
Hara, O., Horinouchi, S., Uozumi, T. and Beppu, T., 1983, J. Gen. Microbiol., 129:2939-2944.
Hopwood, D.A. and Wright, H.M., 1983, J. Gen. Microbiol, 129:3573-3579.
Horinouchi, S. and Beppu, T., 1984, Agric. Biol. Chem., 48:2131-2133.
Horinouchi, S. and Beppu, T., 1985, J. Bacteriol., in press.
Horinouchi, S., Hara, O. and Beppu, T., 1983, J. Bacteriol., 155:1238-1248.
Horinouchi, S., Kumada, U. and Beppu, T., 1984, J. Bacteriol., 158:481-487.
Horinouchi, S., Nishiyama, M., Suzuki, H., Kumada, Y. and Beppu, T., 1985, J. Antibiot., in press.
Horinouchi, S., Suzuki, H. and Beppu, T., 1985, submitted for publication.
Khokhlov, A.S., 1980, in: "Frontiers of Bioorganic Chemistry and Molecular Biology", S.N. Ananchenko, ed., Pergamon Press, Oxford and New York, pp 201-210.
Khokhlov, A.S., Anisova, L.N., Tovarova, I.I., Kleiner, F.M., Kovalenko, I.V., Krasilnikova, O.I., Konitskaya, E.Ya. and Pliner, S.A., 1973, Z. Allg. Mikrobiol., 13:647-655.
Khokhlov, A.S., Tovarova, I.I., Borisova, L.N., Pliner, S.A., Shevchenko, L.A., Kornitskaya, E.Ya., Ivkina, N.S. and Rapport, I.A., 1967, Doklady AN SSSR, 177:232-235.
Kirby, R. and Hopwood, D.A., 1977, J. Gen. Microbiol., 98:239-252.
Lakey, J.H., Lea, E.J.A., Rudd, B.A.M., Wright, H.M. and Hopwood, D.A., 1983, J. Gen. Microbiol., 129:3565-3573.
Mori, K., 1983, Tetrahedron, 39:3107-3109.
Rosenberg, M., Nd Court, D., 1979, Annu. Rev. Genet., 13:319-353.
Rudd, B.A.M. and Hopwood, D.A., 1979, J. Gen. Microbiol., 114:35-43.
Rudd, B.A.M. and Hopwood, D.A., 1980, J. Gen. Microbiol., 119:333-340.
Shine, J. and Dalgarno, L., 1974, Proc. Natl. Acad. Sci. USA, 71:1342-1346.
Thompson, C.J. and Gray, G.S., 1983, Proc. Natl. Acad. Sci. USA, 80:5190-5194.
Westpheling, J., Ranes, M. and Losick, R., 1985, Nature, 313:22-25.
Wright, L.F. and Hopwood, D.A., 1976, J. Gen. Microbiol., 95:96-106.

APPLICATION OF RECOMBINANT DNA TECHNOLOGY IN ANTIBIOTIC-PRODUCING

MICROORGANISMS

G. Holt, P. Ford, Z. Ikram, T.M. Picknett and G. Saunders

The School of Biotechnology, The Polytechnic of Central London
115 New Cavendish Street, London W1M 8JS, UK

INTRODUCTION

A large part of biotechnology is concerned with the discovery and subsequent optimisation of useful biological processes or products (Bull et al., 1982). Optimisation will include complex fermentation and process design in order to produce and purify products in the most effective manner possible. This part of development may be considered the realm of the biochemical engineer and microbial physiologist. A significant input in development however comes about as a consequence of genetic manipulation or alteration of the microbe involved. This can include changes in growth characteristics to suit process design, improvement in yield of product and changes in product so as to improve stability or biological activity (Rowlands and Normansell, 1983) as well as the discovery of novelty. Examples of all of these types of strain development are to be found within the area of antibiotic production.

Antibiotics are structurally and functionally diverse compounds produced in the main by Actinomycetes, particularly Streptomycetes, filamentous fungi and other bacteria including species of Bacillus and Pseudomonas (Berdy, 1974). Traditionally high yielding antibiotic producers are not found in nature and yield improvement immediately becomes of utmost importance in the realization of a pharmacologically proven compound as a commercially produced antibiotic. Over the past 40 years this has success-fully been achieved for many different antibiotic fermentations using what many would term an empirical approach, employing mutation (with radiations and chemicals) of whole cells, sometimes in concert with selection. This process is termed empirical in that very little if any effort was made to determine the biochemical or regulatory basis of the increase in yield obtained. Having said this, such an uninformed approach should never be treated with scorn since its success cannot be disputed nor can it be realistically stated that its use in the near future is likely to diminish to any great extent. There will always be circumstances where random mutation and screening will be necessary either due to its unarguable ability to produce 'quick' and hence economically favourable results or due to the lack of any tractable genetic system in any newly discovered organism.

It is useful to imagine that a major contributing factor to the almost exclusive use of mutation and selection/screening methods in antibiotic production was due to the fact that there really was no alternative genetic technique for bringing about such dramatic changes, certainly not one, which by virtue of its nature, automatically confers on the operator an unparalleled degree of control over the manipulation taking place. Genetic engineering is a technique which can just as easily be used to dissect and analyse a genetic change as to bring it about in the first place, supplying the investigator with a new set of tools with which to establish gene structure, regulation and genome organization.

In order to apply this new approach to genetic improvement of antibiotic producing species it is first necessary to develop suitable cloning vectors. This article will review the vectors which are available within the major antibiotic producing groups and account for the progress to date in the cloning of genes involved in antibiotic biosynthesis.

GENE CLONING IN STREPTOMYCES

Table 1 lists reported cloning vectors for a range of Streptomyces species many of which are antibiotic producers. Vector construction for Streptomyces has been aided considerably by the presence in a large number of strains of plasmid DNA. For example in a survey of wild-type isolates Daniel and Tiraby (1982) detected plasmid DNA in 21 out of 120 isolates tested. Other observers put the frequency of plasmid detection as high as 1 in 3. Kirby et al., (1982) list 10 antibiotic producing species containing plasmid DNA. The majority of these plasmids elicit a 'lethal zygosis' effect manifested by the appearance of a zone of inhibition on a lawn of plasmid-free growth by colonies harbouring plasmid (Bibb et al., 1977). This effect was of considerable use in the initial development of transformation systems in Streptomyces (Thompson et al., 1982; Chater et al., 1982).

In the past it was a commonly held view that genes for antibiotic production in streptomycetes might be found on plasmids and much indirect evidence based on the use of plasmid curing agents was accumulated (Kirby, 1978). Although reports on the involvement of plasmid elements in anti-biotic production continue to appear (see for example Zippel et al., 1983) in only one case, that of methylenomycin A, has it clearly been demonstrated that antibiotic biosynthetic genes are present on an autonomously repli-cating plasmid (Kirby et al., 1975).

The plasmids available, together with antibiotic resistance genes found in antibiotic-producing strains, have successfully been used to construct a range of plasmid-derived vectors with a number of different characteristics. Such constructions were pioneered by the group led by David Hopwood at the John Innes Institute, England. There are currently available several fine reviews dealing with the work of this group (Keiser et al., 1982; Bibb et al., 1983; Hopwood et al., 1986) and readers are directed to these for further detailed information. Briefly, three plasmids, SCP2* (Bibb and Hopwood, 1981), SLP1.2 (Bibb et al., 1980) and pIJ 01 (Keiser et al., 1982) have been developed by these workers. Numerous genes for antibiotic resistance have been identified in Streptomyces and such antibiotic resistance genes have been inserted into these plasmids yielding vectors which can be positively selected for and used to generate recombinants via insertional inactivation. For example pIJ364 carries genes coding for resistance to viomycin and thiostrepton. The thiostrepton gene has a single ClaI site which can be used for insertional inactivation purposes (Keiser et al., 1982). Similarly pIJ702 (Katz et al., 1983) is a high copy number, broad host range plasmid which combines thiostrepton resistance with a gene

Table 1. Cloning Vectors for a Range of Streptomyces Species

Streptomyces species	Vector	Comments	Reference
S. ambofaciens	pFJ103	Bifunctional vector based on plasmid from S. granuloruber and pBR322 carries neomycin and thiostrepton resistance. Host range includes S. fradiae and S. aureofaciens.	Richardson et al., 1982
S. ambofaciens	pKC356	Capable of replication in E. coli and S. ambofaciens. Carries a promoterless transposon Tn5 neomycin resistance gene and has been used to demonstrate that the promoters p-Nm and P1 can function in S. ambofaciens.	Kuhstross et al., 1985
	pKC388	Contains p-Nm inserted upstream from the pKC356 neomycin gene. Plasmids pKC356 and pKC388 used for the characterization of promoters and terminators in S. ambofaciens.	
S. ambofaciens	–	Plasmids constructed using the Bacillus subtilis veg promoter to drive expression of heterologous genes in S. ambofaciens. Transformants were selected for expression of an E. coli hygromycin phosphotransferase gene.	Richardson et al., 1985
S. antibioticus 326	–	Hybrid plasmids based on pBR325 containing extrachromosomal DNA, cSA1, present as a multiple tandemly repeated sequence. Used for the construction of an integrative amplifying vector in S. antibioticus 326	Danilenko and Orlova, 1983
S. azureus	SAt1	Phage vector SAtI is inducible with mitomycin C and UV, multiplies easily with high yields and has thermostable characteristics. Approx. 24 MD in size.	Ogata et al., 1985
S. cattleya	TG1	Broad host range phage vector. TG1 DNA is double stranded, 41 kb in length and has unique sites for ClaI, NdeI, PstI, SmaI and XbaI.	Foor et al., 1985
S. chartreusis	SF1623	Range of cloning vectors constructed using a pock forming plasmid and cloning genes coding for resistence to Kanamycin novobiocin, destomycin and racemomycin.	Murakami et al., 1983
S. fradiae	PJL197	E. coli/Streptomyces Shuttle vector selected positively in both hosts. Recombinant of SCP2* and pBR325. Host range includes – S. griseofulvum, S. fradiae, S. ambofaciens. Single sites in neomycin and thiostrepton resistance genes.	Larson and Hershberger, 1984
S. kasugaensis	oSJ21-B5	Constructed from pSK1 plasmid from S. kasugaensis and thiostrepton resistance gene. Single sites for BclI and SalI. Wide host range for antibiotic producing Streptomyces.	Nabeshima et al., 1984
S. lavendulae	pkST2	Pock forming plasmid carrying genes for resistance to thiostrepton and Streptothricin.	Kobayashi et al., 1984
S. lividans	pSLP124	Promoter probe vector. Based on E. coli chloramphenical acetyltransferase gene from pACYC184 lacking DNA sequence normally recognised by RNA polymerase.	Bibb and Cohen, 1982
S. lividans	p1J424	Promoter probe vector constructed by insertion of E. coli kanamycin resistance determinant Tn5 into p1J702. Expression of Tn5 is normally prevented in this vector by the insertion of a phage terminator between the promoter and structural gene.	Bibb, M., cited in Martin and Gil, 1984
S. lividans	pBGH007	S. lividans expression bector pBGH007 contains the bovine growth hormone gene under control of the S. fradiae aminoglycoside 3'-phosphotransferase regulatory region.	Gray et al, 1984
S. lividans	pARC1	Promoter-probe vector constructed from p1J41 containing gene(s) directing the synthesis of a brown pigment. The E. coli lac (trp-lac hybrid) promoter ligated into pARC1 promoted expression of cloned pigment gene(s) in S. lividans.	Horinouchi and Beppu, 1985

Table 1. Cloning Vectors for a Range of Streptomyces Species (Cont'd)

Streptomyces species	Vector	Comments	Reference
S. lividans	pFSH102	Shuttle vector for Streptomyces spp. and E. coli expressing sulphonamide resistance in both genera.	Shareck et al., 1984
S. lividans	pMT660	Temperature-sensitive plasmid replication mutant of plJ702. May be used in the detection of transposable elements in replicon fusions of Streptomyces DNA, or as an alternative to phage ØC31-based vectors in mutational cloning systems.	Birch and Cullum, 1985
S. lividans	pBV3 pBV4	Hybrid plasmids derived from plJ702, encoding a streptomycin phosphotrans-ferase. Selection on thiostrepton and streptomycin.	Vallins and Baumberg, 1985
S. parvulus	R4 22B tsr-1	Deletion derivative of temperate phage R4 carries thiostrepton resistance gene.	Morino et al., 1984
S. rimosus	pSB24.1.	Composite plasmids derived from SLP.2 and multicopy plasmid from S. cyanogenus. Able to amplify kanamycin resistance of S. rimosus to 50 mg/ml. Single sites for BamHI, ClaI and SacI.	Danilenko et al., 1984
S. rimosus	pPZ12	Deletion variant of plJ303 obtained from S. rimosus transformants. Single sites for PstI, ClaI, KprI and SacI.	Rhodes et al., 1984
S. rimosus	pPZ74	Bifunctional cosmid vector composed of plJ303 and pBR325 sequences plus λ cos sequences. Capable of accomodating 25-40 kb inserts.	Chambers and Hunter 1984
S. roseochromogenes	pSRCIb	Pock forming plasmid carrying thiostrepton resistance. Wide host range including S. lavendulae, S. parvulus and S. actuosus. Single site for BglII.	Shindoh et al., 1984
Streptomyces spp.	pUC1116	Bifunctional shuttle vector expressing resistance to ampicillin, chloramphenicol and viomycin in E. coli, thiostrepton, viomycin and the enzyme tyrosinase in streptomyces species.	Manis and Clemens, 1984
Streptomyces spp.	pFJ269	Plasmid derived from pFJ278 encoding thio-strepton resistance. Copy number - about 40. Broad host range with a transformation frequency of 1000-10,000/μg. Unique restriction sites for ClaI and PstI.	Jones and Fayerman, 1984
	pF301	pFJ276/pBR328 hybrid vector contains unique sites for ECORl, ClaI and SacI.	
Streptomyces spp.	KC505 KC515 KC516 KC518	Phage vectors derived from ØC31 carrying resistance to thiostrepton and a cloned promoterless viomycin-phosphotransferase gene. Capable of accomodating 4 kb inserts. KC505 is a PstI replacement vector. KC518 is a replacement vector for DNA fragments with 5'-GATC ends.	Rodicio et al, 1985

encoding the production of tyrosinase. Cells harbouring plJ702 are pheno-
typically thiostrepton resistant and appear black on medium containing
tyrosine, due to the tyrosinase-directed synthesis of melanin. The tyrosinase gene has unique enzyme sites for BglII, SphI and S tI. Inserts at
these sites interrupt the tyrosinase gene giving colonies which are
thiostrepton resistant but white in appearance.

Streptomycete phages have also been developed as cloning vectors at the
John Innes: derivatives of the temperate phage ØC31, most particularly
ØC31:KC400 which carries viomycin resistance, can be used in the process
termed mutational cloning (Hopwood et al., 1983) the implications of which
will be discussed in greater detail below. Recent reports have demonstrated
the development of two other broad host range temperate phage vector based
on R4 (Morino et al., 1984) and TG1 (Foor et al., 1985).

Several new types of vector have also been developed, including
promoter probe vectors such as pARC1 (Horinonchi and Beppu, 1985), a plasmid
that allows chromogenic identification in S. lividans as well as expression
vectors such as pBGH007 (Gray et al., 1984), designed to permit production
of proteins from Bacillus and E. coli in Streptomyces spp. (see Table 1). A
temperature sensitive plasmid vector pMT660 (Birch and Cullum, 1985) may
also be useful as a vector for transposon mutagenesis and as an alternative
to phage ØC31-based vectors used in mutational cloning systems. Another
interesting development of potential importance has been that of an
integrative amplifying vector in S. antibioticus 326 containing a multiple
tandemly repeated sequence existing in multiple copies in S. antibioticus
1607 (Danilenko and Orlova, 1983). This may be of significant industrial
importance as it can be used to amplify cloned genes.

The number of plasmid and phage-based cloning vectors is thus steadily
increasing (see Table 1) and the appearance of such reports is, together
with numerous patents for pock-forming plasmids, becoming quite commonplace.
Not surprisingly, it is now possible to assemble an impressive list of
cloned Streptomyces genes, some of which are involved in antibiotic
biosynthesis (see Table 2 and below).

Examples of Gene Cloning in Streptomycetes

There are a number of reasons for cloning genes from Streptomyces
stemming from their almost ubiquitous production of antibiotics and the
complexity of this process in general. Resistance genes are of importance
initially in the actual construction of cloning vectors but they may also
have other uses. As pointed out by Thompson and Davies (1984), many high-
yielding antibiotic producing strains rely on the fact that they are
resistant to the high concentrations of antibiotic which they produce
(Thompson et al., 1982). Indeed in some cases it has proved possible to
isolate higher producing strains by virtue of their elevated resistance to
the antibiotic produced (See Holt and Saunders, 1985). If as hoped the
future sees a significant elevation in antibiotic titres due to gene
amplification it may become necessary to amplify resistance at the same time
or to engineer in one strain more than one resistance mechanism.

In addition, the cloning of antibiotic resistance genes has led
indirectly to the identification of gene sequences governing the biosynthesis of the antibiotic concerned (Stanzate et al., 1986). Such genes have
also furnished geneticists working in this general area with some highly
active promoters as evidenced, in one particular study, by the presence of
aminoglycoside phosphotransferase as 10% of the total soluble protein
(Thompson et al., 1982). Such high activity, combined with a high copy
number plasmid (Keiser et al., 1982) could be of considerable value bearing

Table 2.   Examples of the Isolation of Genes from _Streptomyces_ Species

Organism	Gene cloned	Comments	Reference
S. alboniger	Puromycin N-acetyl transferase	Vector used pIJ702. Puromycin N-acetyl transferase gene cloned and expressed in S. lividans and E. coli.	Vara et al, 1985
S. antibioticus	Tyrosinase	Vector used pIJ37. Colonies secreting tyrosinase appear black. Used for insertional activation purposes in vector pIJ702.	Katz et al, 1983
S. antibioticus	Phenoxazinone-synthase	Vector used pIJ702. Transformants screened for ability to produce phenoxazinone synthase, involved in actinomycin biosynthesis.	Jones and Hopwood, 1984a,b.
S. bikiniensis	Regulatory and structural genes for A factor (2-isocapryloyl-3R-hydroxymethyl-γ butyrolactone)	Vector used pIJ385. Clone detected by restoration of ability to synthesise streptomycin	Horinouchi et al, 1984
S. clavuligerus	Gene involved in biosynthesis of clavulanic acid	Vector used pIJ702. Gene library first constructed using S. lividans as host. Clone detected by complementation of npe mutant.	Bailey et al, 1984
S. coelicolor	hisD3., argA1, guaA1	Primary biosynthetic genes cloned using SLP1.2. Clones detected by complementation of auxotrophy.	Thompson et al, 1982
S. coelicolor	Agarase	Vector used pIJ702. High level (500 times greater) expression of agarase obtained in S. lividans. Potentially useful to export cloned gene products across cell membrane/wall.	Kendall and Cullum, 1984
S. coelicolor	Genes for glycerol-3 phosphate dehydrogenase and glycerol kinase	Vector used pIJ702. Investigation of the gyl operon is being undertaken using mutational cloning. gyl operon subject to glucose repression and of interest from this point of view.	Seno et al., 1984
S. griseus	Gene coding for paba synthetase	Vector used pIJ41. Clones selected on basis of resistance to sulphanamide and relief of paba auxotrophy in S. lividans.	Gil and Hopwood, 1983
S. griseus	Streptomycin 6-0-phosphotransferase	Vector used pIJ702. Resistance to streptomycin expressed in S. lividans.	Tohyama et al, 1984
S. griseus	Streptomycin phosphotransferase	Vector used pIJ702. Transformants of S. lividans selected for both thiostrepton and streptomycin resistance.	Vallins and Baumberg, 1985
S. griseus	Str R, str A, str B and str C. Streptomycin gene cluster	Vector used pOA154. An str gene cluster containing at least 4 genes involved in streptomycin biosynthesis, including gene for streptomycin-6-phosphotransferase, amidinotransferase.	Ohnuki, 1985
S. hygroscopicus	Hygromycin B[R]	Vector used pIJ350. Resistance expressed in S. lividans.	Malpartida et al., 1983
S. lividans	β-galactosidase like activity	Vector used pIJ6. Colonies carrying recombinant plasmid give blue halo on plates containing X-gal.	Taylor et al., 1983
S. rimosus forma paromomycinus	Paromomycin-phosphotransferase (pph)	Vector used pIJ702. Selection of S. lividans transformants on paromomycin.	Perez-Gonzalez and Jimenez, 1984
S. vinaceous	Viomycin resistance determinate	Vector used SLP1.2. Viomycin since used to tag family of vectors derived from pIJ101.	Thompson et al, 1982

in mind the practical aims of cloning in improving antibiotic synthesis. In
a similar manner the cloning and expression of genes for extra-cellular
enzymes, such as the β-galactosidase of S. lividans (Taylor et al., 1983)
and the agarase enzyme of S. coelicolor (Kendall and Cullum, 1984), have in
addition to intrinsic interest in their structure and regulation, a
potential value in the export of foreign proteins expressed in streptomycetes across cell membrane and wall.

    Cloning of genes in primary metabolism is important since primary
metabolic pathways provide the precursors for antibiotic structures.
Therefore manipulation of primary pathways by genetic engineering may be
just as important in elevating antibiotic titre as changes in the enzymes
actually concerned with assembling the sub-units of the antibiotic. In this
respect the cloning and analysis of the glycerol utilisation operon (gyl) of
S. coelicolor is of interest (Seno et al., 1984). Growth of this organism
on glycerol as sole carbon source depends upon the enzymes glycerol kinase
(gylA) and glycerol-3-phosphate dehydrogenase (gylB) (Seno and Chater,
1983). Using the plasmid vector pIJ702, these authors cloned, as it turned
out, promoterless sequences which could complement gylA and gylB mutants.
Subsequently the mutational cloning phage vector ØC31:KC400 was used
successfully in gene disruption experiments on the gyl operon.

    Using this technique, it was possible to assign particular functions to
specific domains of the operon and obtain evidence to the fact that the
gylA, gylB genes are transcribed as a polycistronic message. Cloned gyl DNA
has been used as a probe and led to the cloning of gyl DNA from S. griseus,
a commercial antibiotic producer (Bibb et al., 1983). This will allow study
of this catabolic operon in an organism of industrial importance.

    The cloning of genes directly involved in assembling an antibiotic
should eventually provide answers to many questions. For example, do genes
occur in typical operon type units, are there other specific regulatory
signals in addition to the more general modes already established such as
catabolite repression? A fine start to the cloning of antibiotic biosynthetic genes has been made once again by workers at the John Innes
Institute. Many general reviews are available on their work in this
area (e.g. Hopwood et al., 1983) to which the reader is directed.

    Of special interest are the means available with which to detect
recombinant plasmids carrying cloned copies of antibiotic pathway genes.
The most commonly used technique is complementation of antibiotic non-
producing (Npe) derivatives with restoration of antibiotic production and
bioactivity as the criteria for success. In one case, the cloning of the
para-amino benzoic acid synthetase gene (paba) involved in the biosynthesis
of candicidin in Streptomyces griseus (Gil and Hopwood, 1983), it was
possible to select clones directly. Donor DNA was used from a paba
overproducing mutant which can be selected for on the basis of resistance to
sulphonamide. Further studies with the cloned paba synthetase gene has
resulted in the isolation of its presumed promoter sequence using the
promoter probe vector PIJ414 (Gil unpublished cited in Martin and Gil,
1984). It was found that the paba synthetase gene could be expressed in
E. coli but only after removal of the presumed promoter sequence of the
Streptomycete gene, expression occurring by transcriptional readthrough from
the tetracycline resistance gene of pBR322 (Bibb et al., 1983). In another
variation on this theme, Feitelsen and Hopwood (1983) have cloned an 0-
methyl-transferase involved in the biosynthesis of undecyl prodigiosin where
clone identification was facilitated by the pigmented nature of the
antibiotic. Chater has developed an intriguing technique for the cloning of
antibiotic production genes based on the process termed mutational cloning
using the att defective ØC31 phage derivate ØC31:KC400 (see Chater and

Bruton, 1985; Bibb et al., 1983; Hopwood et al., 1983). The phage carries viomycin resistance, can accomodate 2-6Kb Pst I fragments and transduces recipients by recombination/integration between cloned sequences in the phage and homologous sequences on the chromosome. This technique has already been used to isolate 8Kb of DNA coding for genes involved in methylenomycin biosynthesis, which were known to be carried on the S. coelicolor plasmid SCP1 (Kirby and Hopwood, 1977). DNA from an S. parvulus strain carrying SCP1 was shotgun cloned into ØC31:KC400 and the resulting ligation mix used to lysogenize S. lividans or S. coelicolor strains carrying an integrated copy of SCP1. The SCP1 sequences should provide the only sites of homology between the S. parvulus library and the two recipients. A feature of the use of ØC31:KC400 is that under certain circumstances, when the cloned fragment lacks promoter and C-terminal sequences, integration into the wild type gene can result in disruption and loss of function (Hopwood et al., 1983). With antibiotics, non-production is the result and in the case discussed above of 278 ØC31:KC400 clones carrying SCP1 fragments, used to transduce an S. coelicolor strain with an integrated copy of SCP1, 9 non-producers generated by integration were detected. This process may lend itself to general use in circumstances where suitable npe strains are not available, being limited only by the host-range of ØC31, the presence of restriction barriers within strains and indigenous resistance to viomycin.

Two other examples of the cloning of genes involved in antibiotic synthesis are of interest. One involves the production of the β-lactam antibiotic clavulanic acid and illustrates the use of mutants blocked in the biosynthetic pathway. Baily et al., (1984) have utilised a non-clavulanic acid producing mutant of S. clavuligerus as recipient for a gene library of wild type DNA from this organism constructed in pIJ702 using S. lividans as host. Clones carrying DNA complementary to the dcl-8 mutation were detected by screening transformants using a plate assay for the β-lactamase inhibitory activity of clavulanic acid.

The second example, that of A-factor (2-isocapryloyl-3R-hydroxymethyl-8-butyrolactone) is fully described in the article by Bepput (1986) in this volume.

GENE CLONING IN OTHER ACTINOMYCETES, BACILLUS AND PSEUDOMONA

The clear effort which has gone into developing vector systems for species of Streptomyces has not been mirrored for other Actinomycetes. This undoubtedly reflects the fact that the streptomycetes produce by far the greater majority of antibiotics within the filamentous bacterial group and are usually the most prevalent type found in soil (Taylor et al., 1983). However, the same potential for genetic manipulation exists for other Actinomycetes and a few reports of the technology required for cloning are emerging. Taylor et al. (1983) have reported initial investigations into protoplasting techniques and the presence of a plasmid pSgB-1 in Strepto-sporanguin brasiliense. However, the authors report failure in all their attempts to transform this organism.

Excellent cloning systems are however available for the minor anti-biotic producing genera Bacillus and Pseudomonas, and it is perhaps surprising, in view of this fact, that there are not many reports in the literature of the use of cloning to improve or study antibiotic synthesis in either. Species of both genera produce a variety of antibiotic substances, some of which are listed in Tables 3 and 4. Species of bacilli produce mainly peptide antibiotics, B. subtilus being the major producer (approximately 65-70 different peptides reported), followed by B. brevis

Table 3. Some Antibiotics Produced by the Genus Bacillus

Antibiotic	Species	Reference
Bacillomycin F	B. subtilis	Peypoux, 1985
Bacitracin	B. subtilis	Zimmer et al., 1979
Bmy-28160	B. circulans	Sugawara et al., 1984
Butirosin	B. circulans	Herbert et al., 1986
Colistin	B. collistinus	Shirai et al., 1984
Gramicidin S	B. brevis	Izumiya et al., 1979
Megacin	B. megaterium	von Tersch and Carlton, 1983
Mycobacillin	B. subtilus	Majumder et al., 1985
Polymyxin	B. polymyxa	Storm et al., 1977
Proticin	B. licheniformis	Katz and Demain, 1977
Subtilin	B. subtilus	Nishio et al., 1983
Tyrocidines	B. brevis	Izumiya et al., 1979

(20-25 different antibiotics). An excellent review on antibiotics produced by Bacillus has been written by Katz and Demain (1977).

In bacilli, vector development has reached a stage where it can reasonably be said that the available expertise is second only to that established for E. coli. An almost continuous stream of reports of vector development has appeared in the published literature, a situation undoubtedly prompted by the industrial potential of bacilli, in areas other than antibiotic synthesis. Bacillus species are widely used in the fermentation industry to produce a variety of economically important enzymes, amino-acids, nucleotides and vitamins. In addition to this, their ability to grow to high cell densities, and their acceptability for food products make them an attractive choice for future production of foreign proteins.

Table 4. Some Antibiotics Produced by Pseudomonas Species

Antibiotic	Species	Reference
Cepacin A & B	P. cepacia	Parker et al., 1984
Obafluorin	P. fluorescens	Wells et al., 1984
Pyrrolnitrin	P. flourescens	Elander, 1982
Sulfazecin and Isosulfazecin	P. Acidophila	Demain, 1983
Tabtoxin	P. tabaci	Demain et al., 1983
Thiotropocin	P. CB-104	Kintak et al., 1984

Although many species of Bacillus harbour their own plasmids, most of
them do not express identifiable genetic traits. However, a B. cereus
plasmid pBC16, encoding tetracycline resistance, has been used to construct
a variety of cloning vectors. Antibiotic-producing strains of B. megaterium
have been analysed for resident plasmids, and one strain was discovered to
contain at least ten plasmid sizes, one of which was found to code for the
antibiotic megacin (von Tersch and Carlton, 1983). Another strain had at
least seven stable plasmids, present in high copy number, thus reflecting
the possibilities of cloning foreign DNA into this species (Kieselburg et
al., 1984).

The majority of vectors available for cloning in bacilli are derived
from plasmids originally isolated from S. aureus and have a broad host range
within the genus (Ehrlich, 1977; Gryczan et al., 1978). A number of useful
chimeric vectors have also been constructed, capable of replicating in
B. subtilus and E. coli. The majority have been derived by ligation of
pC194, from S. aureus and pBR322, from E. coli (Ehrlich, 1978; Gryczan and
Dubanau, 1978). More recently shuttle vectors derived from runaway
replication plasmids, related to CloDF13 (a bactericinogenic plasmid), have
been reported (Andreoli, 1985). Such vectors can be amplified in E. coli,
to high levels by elevating the temperature and hence could prove extremely
useful for production of large quantities of DNA and encoded proteins.

Reported applications of gene cloning techniques to antibiotic
producing species of Bacillus are beginning to appear in the literature.
Hoshino et al. (1984) have used the shuttle vector pTA1302 to clone a DNA
fragment encoding resistance to butirosin and other aminoglycoside
antibiotics, derived from the chromosome of B. vitellinus 2-1159, a producer
of butirosin. The gene has been cloned and expressed in both E. coli and
B. subtilus. An ornithine-activating fragment of the Gramicidin 5
synthetase 2 gene has also been cloned from B. brevis (Krause et al., 1985).

The application of recombinant DNA techniques to the study of antibiotic biosynthesis would be greatly facilitated by parallel biochemical
investigations of the pathway of interest. In this connection the use of
"blocked mutants" has proved very effective in the study of secondary metabolic pathways. For example, in B. brevis blocked mutants have resulted in
the isolation and characterization of some peptides possibly involved in
mycobacillin biosynthesis, and have facilitated the location of enzyme
defects in the mycobacillin synthetase complex (Majumder et al., 1985).
Such mutants could undoubtedly be used to clone genes involved in
mycobacillin synthesis.

Many members of the genus Pseudomonas elicit marked nutritional and
biochemical versatility, and are also easily handled under laboratory
conditions. Pseudomonas spp. are a source of many interesting compounds,
for example enzymes such as proteases and lipases, phenzines, quinoline
derivatives, keto acids, vitamin B_{12}, enzyme inhibitors and antibiotics. A
number of antibiotics produced by this genus are listed in Table 4. From
the genetic viewpoint, P. aeruginosa is the best known member of the
species. However, genetic understanding of P. putida has rapidly developed
over the past couple of years (Holloway et al., 1979).

Vectors commonly used for E. coli and B. subtilus cannot be used for
cloning in Pseudomonas, showing marked instability upon introduction into
members of this species. A versatile cloning vector, especially for
Pseudomonas species was reported by Wood et al., (1981) to overcome this
difficulty. In addition, a broad-host-range vector has been reported by
Olsen et al., (1982). They developed a host-vector system for P. aeruginosa
PAO by cloning scattered regions of the PAO chromosome from a DNA gene bank.

The progenitor of the vector was a small plasmid pRO1600 found in a PAO strain which had acquired RPI in a mating experiment. RPI is very similar to RP4 and RK2, the wide-range plasmids of IncP-1 (Incompatibility-1) group. These plasmids can be transferred to members of many Gram-negative genera. Unfortunately, most broad-host-range plasmids contain few restriction endonuclease cleavage sites and even fewer unique sites. The broad-host-range vector RSF1010 has no sites for Hind III, BamHI, Bgl II, Sal I, xba I, Xho I, Xma I, and Kpu I endonucleases. In view of this fact, the cloning and expression of genes in Pseudomonas has been difficult and limited.

Recently several broad-host-range shuttle vectors have been constructed (Werneke et al., 1985). These vectors are capable of expression in Pseudomonas or E. coli and possess easily selectable antibiotic resistance markers with multiple cloning sites. These vectors were constructed by insertion of the entire pUC13 sequence into derivatives of RSF1010. Another E. coli - Pseudomonas shuttle vector reported for cloning and characterization of Pseudomonas genes, employed a transposable element, Tn3-delta-596 (Kok et al., 1984). Thus for Pseudomonas species there now exist plasmids which can be used to shuttle genes between this genus and E. coli. Although Pseudomonas genes are not expressed very well in E. coli, Matsuda and Komatsu (1985) have reported the cloning of 7β-(4-carboxy-butanamido) cephalosporanic acid acylase from a Pseudomonas strain, in E. coli. An enzyme capable of deacylating 7β-(4-carboxybutanamido) cephalosporanic acid (Glutarylamido cephalosporanic acid, GL-7AC) to 7-aminocephalosporanic acid (7-ACA) is found in a few strains of Pseudomonas. Much interest has developed towards the properties of this enzyme, and its possible application to an industrial process for 7-ACA preparation. Despite the multicopy nature of plasmid pBR325, on which the gene for this enzyme was cloned in E. coli, Matsuda and Komatsu (1985) found that the enzyme activity was very low. However, when they reduced the size of the initial cloned fragment, and placed it at a region downstream of the E. coli promoter, a significant increase in enzyme activity was observed.

Like many other examples, analogue resistant mutants have successfully been used to increase the antibiotic production in Pseudomonas for those antibiotics which have amino acids as precursors. Selection of higher pyrrolnitrin producing strains of P. flourescens was reported by Elander (1982). Pyrrolnitrin is an important antifungal, for which D-tryptophan is a precursor. Flouro or methyl tryptophan added to the media generated a set of mutants that overproduce D-tryptophan, resulting in high levels of pyrrolnitrin production, also eliminating the need to add D-trytophan to the growth medium. As for species of Bacillus there are at present few reports of the use of genetics in the study or production of antibiotics by species of Pseudomonas.

CLONING IN THE FILAMENTOUS FUNGI

Neurospora crassa and Aspergillus nidulans

Cloning systems for filamentous fungi are not as far advanced as those available for streptomycetes and other procaryotes. According to Berdy (1974) filamentous fungi produce a total of 868 antibiotics (10 of which are commercially utilised), second only to the Actinomycetes in terms of total number. The major efforts in vector construction have focussed to date on two members of the Ascomycotina, Neurospora crassa and Aspergillus nidulans. This is not surprising in view of the long history of classical genetic studies associated with these two species (see Mishra, 1985; Turner and Ballance, 1985). Clearly further development of host vector systems for the filamentous fungi, particularly with antibiotic producing species, (see below) is essential in order to generate the same possibilities now being realised for the streptomycetes.

At present there are two widely felt limitations hampering efforts to develop broad host range vectors for filamentous fungi (Saunders et al., 1986). The first involves an almost complete lack of suitable endogenous plasmids or phages on which to base the construction of cloning vehicles. Plasmids of mitochondrial origin do occur but these are of limited value especially in circumstances where appearance of plasmid is associated with a particular degenerate phenotype (Tudzynski and Esser, 1985). The advantages afforded by the availability of a cytosplasmically located plasmid are evidenced by the speed with which autonomously replicating yeast vectors, based on the 2μm plasmid, were developed (Hollenberg, 1982). The second general limitation stems from a lack of suitable, dominant resistance markers, which can be cloned and used to effectively select for trans-formants. Although advances have been made with the use of Escherichia coli resistance elements such as Tn:902, conferring resistance to G418 (Jiminez and Davies, 1980), their use has not yet become widespread, possibly due to the general insensitivity of fungal species to such agents. One vector has however been constructed for the filamentous species Ustilago maydis based upon selection for resistance to the antibiotic neomycin. Transformation frequencies in this system are low (1-10 transformants per μg DNA) (Banks, G., personal communication).

Despite these limitations several groups have successfully developed host vector systems for a number of filamentous fungi (see Table 5 and Saunders et al., 1986). None of these vectors have a broad host range as each is detected in transformants by the relief of specific biosynthetic deficiencies. In the case of the first reported vector for filamentous fungi, pVK88 of N. crassa (Case et al., 1979) the gene coding for an enzyme involved in quinic acid catabolism, catabolic dehydroquinase (CDQ) was cloned in the following manner. CDQ is an isoenzyme of biosynthetic dehydroquinase (BDQ), an enzyme involved in aromatic amino acid biosynthesis in both N. crassa and E. coli. Thus CDQ from N. crassa should complement BDQ deficient E. coli (aroD) mutants allowing the cloning of N. crassa CDQ on an E. coli plasmid vector (Vapnek 1977). By isolating this recombinant plasmid from E. coli, N. crassa mutants lacking both CDQ and BDQ could be transformed and selected for on minimal medium lacking an aromatic amino acid supplement. Transformed colonies of N. crassa obtained were found to harbour the recombinant plasmid integrated into the chromosonal DNA of the fungus and subsequently such a vector was referred to as an integrating shuttle vector. The integrative mode of inheritance, although lending stability to cloned genes, makes it difficult to re-isolate DNA for further analysis. As a consequence considerable efforts have been made, with varying degrees of success, to convert such integrative vectors to auto-nomously replicating derivatives. For example, the Labelle mitochondrial plasmid from a wild-type strain of N. crassa (Stohl and Lambowitz, 1983) was inserted into an integrative vector generating the plasmid pALS-1. Low molecular weight DNA, representing free plasmid, was detected in DNA extracts from transformants by Southern hybridisation. Initially it was thought that the conversion to autonomy was caused by the presence of a replication origin (or autonomously replicating sequence ars) in the mitochondrial plasmid. However, a deletion derivative of pALS-1, pALS-2 was subsequently isolated and found to lack possibly all Labelle sequences and yet still to possess autonomous ability. The only advantage conferred by the Labelle sequences in pALS-1 appears to be an enhanced (5 times) trans-formation frequency. The authors' conclusions are not definitive. It still remains unproven that the autonomous function is a consequence of the Labelle sequences. It has been suggested that autonomy may be conferred by a replication origin present either in the bacterial sequences of the replicon or the Neurospora qa $^{2+}$ part used for selection purposes.

Table 5. Examples of Vector Construction for Filamentous Fungi

Organism	Vector	Selection Basis	Comments	Reference
Acremonium chrysogenum	-	Resistance to the amino-glycoside antibiotic G418.	Vector contains Tn903 and a-mitochondrial ars.	Penalva et al., 1985
Aspergillus nidulans	p3 R2	Production of acetamidase. Growth on minimal medium with acetamide as sole nitrogen sources.	Vector integrates at site of acetamidase resident gene. Derivatives containing ribo-repeat and mitochondrial ars do not have improved trans-formation frequency.	Tilburn et al., 1983
	pFB6	Complementation of orotidine-5'-phosphate decarboxylase. Growth on minimal medium lacking uridine.	Vector contains Pyr 4 gene of N. crassa. Integration at more than one site, homology not required.	Ballance et al., 1983
	pSLA43	Complementation of ornithine transcarbamylase deficiency. Growth on minimal medium lacking arginine.	Plasmid can express in yeast and E. coli. Frequency of trans-formation x 10 greater with linearised plasmid.	John and Peberdy, 1984
	pDJB2	Same as for pFB6.	pDJB2 is pFB6 containing a 3.5 kb ars designated ansl selected for in yeast. Trans-formation 100x that of pFB6.	Ballance and Turner, 1985
	pMW11	Resistance to oligomycin	Vector contains oliC31 allele of the A. nidulans oliC gene, isolated by heterologous probing.	Ward et al., 1986
	pPC-3	Complementation of phophoribosyl anthranilate isomerase. Growth on minimal medium lacking tryptophan.	Vector composed of pBR328 carrying the trpC gene from Penicillium chrysogenum	Picknett et al., submitted for publication
Aspergillus niger	p3SR2	As above.	Contains amds gene of A. nidulans. Transformation due to integration of vector in tandem arrays.	Kelly and Hynes, 1985
	pDG3	Production of ornithine transcarbamylase. Growth on minimal medium lacking arginine.	Carries argB gene of A. nidulans. Integrates at various sites in the genome.	Buxton et al., 1985
Cochliobolus Hetero-strophus	p3SR2	As above.	Single or multiple copies of plasmid integrated into chromosome. Transformation frequency 1-3/50μg DNA.	Turgeon et al., 1985
Neurospora crassa	pVK88	Complementation of biosynthetic dehydroquinase. Growth on minimal medium lacking aromatic amino acid supplement.	Linked, unlinked and replacement types of integration occurs.	Case et al, 1979
	pALS1	As for pVK88.	Vector replicates autonomously but also integrates in a functional manner. Incorpor-ation of labelle plasmid in pALS1 improves transformation frequency. pALS2 lacks labelle sequences but still replicates autonomously.	Stohl and Lambowitz, 1983.
	pJR2	Complementation of glutamate dehydrogenase deficiency. Growth on minimal medium supplemented with glycine.	Vector replicates autonomously in deletion host. Plasmid occurs in multimeric form. No recognised ars sequence. Insertion of labelle sequences have no apparent effect.	Grant et al., 1984
Penicillium chrysogenum	pGB83	Complementation of an unspecified auxotrophic mutation.	Transforming plasmid integrates into recipient genome.	Van Solingen et al., 1985
Podospora anserina	pSP17	Expression of senescence. Recipient used is a mutant in which normal senescence is delayed.	Plasmid is autonomous. Expression of bacterial β-lactamase obtained in fungal host after integration into mitochondrial DNA.	Stahl et al., 1982
Ustilago maydis	pMP81	Expression of aminoglyceride phosphotransferase. G418 resistance.	Plasmid autonomous carrying yeast ars.	Banks, 1983

For <u>A. nidulans</u> seven separate vectors have been developed (see Table 5). One of these complements an auxotrophic requirement for uridine in recipient strains (Ballance et al., 1983). The original vector constructed pFB6 carried a gene coding for orotidine-5'-phosphate decarboxylase (<u>pyr</u>4) previously isolated from <u>N. crassa</u>. Using this plasmid Ballance and co-workers transformed <u>A. nidulans</u> at a frequency of 5-10 transformants per µg DNA.

Using yeast selection systems it has proven possible to clone fragments from a range of eukaryotes capable of conferring autonomous replicative ability (<u>ars</u> fragments) on yeast integrative vectors (Stinchcomb et al., 1980). Once isolated and re-introduced into the organism from which it originated such an <u>ars</u> might reasonably be expected to do the same for an integrative vector of that organism. In this way Ballance and Turner isolated a 3.5Kb fragment of chromosomal DNA which possessed the ability to replicate autonomously in yeast. Subcloning this piece of DNA into the integrative vector based on the <u>pyr</u>4 gene of <u>N. crassa</u> resulted in a 100 fold increase in transformation frequency. Optimization of the transformation procedure further improved the frequency so that it is now the highest reported in the filamentous fungi (5×10^3 transformants per µg DNA), further demonstrating that high frequencies of transformation are possible within the filamentous fungi. However transformants obtained in this system still maintain vector DNA in an integrated rather than autonomous state. Although it is encouraging to see already developed a high frequency transformation system the rudimentary nature of filamentous fungal cloning vectors becomes more apparent when it is realized that self cloning, a routine approach in most <u>Streptomyces</u> spp. to the isolation of genes, is only really possible in this one system. Turner and Ballance (1985) have reported the isolation of the <u>acu</u>D gene (coding for isocitrate lyase) in this way. In the main, however, indirect means have to be found to clone genes from filamentous fungi at present (see Saunders et al., 1986). Nevertheless it is encouraging to note that despite this drawback the number of cloned genes from filamentous fungi, particularly <u>A. nidulans</u>, continues to grow at a steady pace. Complete utilisation of the range of molecular biological techniques available, coupled with the development of specialised vectors (for example promoter probe vectors) cited in Turner and Ballance, 1985) has led to the use of <u>A. nidulans</u> as host to produce human interferon (Meetings Report Biotechnology, 4:385-386, 1985).

Penicillium chrysogenum and Acremonium chrysogenum

Within academic institutions comparatively little effort has been applied to developing host/vector systems for antibiotic producing species. Although it is known that <u>A. nidulans</u> produces penicillin and it can be used as a valuable model system, this organism is not at present used commercially. From the point of view of antibiotic production great benefits within the industry will arise as a consequence of the development of high frequency transformation systems specifically for industrially used species such as <u>P. chrysogenum</u> and <u>A. chrysogenum</u>.

Reports of molecular biological studies with these two species are few in number. With <u>A. chrysogenum</u>, Tudzynski and Esser (1982) have reported the isolation of <u>ars</u> sequences in yeast from this organism. In addition, the same group of workers have detected a plasmid-like DNA element in the mitochrondria (Minuth et al., 1982). As regards transformation systems for this organism one gene, complementing the <u>LeuB</u> mutation in <u>E. coli</u> has been cloned (Friedlin and Nuesch, 1984) and one group has reported direct transformation of this organism with selection made on the basis of resistance to G418 (Penalva et al., 1985). Work with this species has also involved the isolation of a gene governing the synthesis of Isopenicillin - by indirect means. In this study sequencing of the isopenicillin N synthetase enzyme permitted the isolation of the gene from a gene bank of A. chrysogenum constructed in E. coli using synthetic oligonucleotide probes synthesised according to predictions made from the protein sequence (Samson et al., 1985).

Work at the Polytechnic of Central London has recently extended classical genetic studies of P. chrysogenum (Normansell et al., 1979) with preliminary investigations into developing the molecular biology of this organism (Saunders and Holt, 1986). Thus the nucleic acid content of this species has been studied in some detail (Saunders et al., 1984a, b; Ford et al., 1986) and ars sequences have been isolated (Ford et al., 1986). In addition three genes capable of complementing proA, leuB and trpC mutations in E. coli have been cloned from this species (Saunders et al., 1986). One other group has already reported transformation of P. chrysogenum although only at a low frequency (van Solingen et al., 1985).

CONCLUSIONS

Recent years have witnessed the transfer of academically established genetic techniques into the practical, applied area of antibiotic production. Technically improved methods of induced mutagenesis, including the use of repair deficient strains (Rowlands and Normansell, 1983; Holt and Saunders, 1985), and recombination via protoplast fusion (Alfoldi, 1982) have allowed a more flexible approach to industrial strain improvement.

Looking to the future, for the complete realization of the potential of gene cloning techniques suitable cloning vectors for use in commercial production strains are required. From the point of view of antibiotic production such vectors have already been constructed for a wide range of species of Streptomyces, Bacillus and Pseudomonas. It appears that at the present time little work involving the application of these vectors to antibiotic production by species of Bacillus or Pseudomonas is ongoing. This situation doubtless reflects the relatively minor commercial significance of these two genera. In contrast, within the Streptomyces, work has advanced at a great pace and as a consequence both plasmids and phage vectors have now been used to clone genes directly involved in antibiotic production, allowing a comparison of their molecular structure and regulation to get underway. Additionally such cloned genes have been used to deregulate antibiotic production (Beppu, 1986) and to generate novel antibiotic structures (Hopwood et al., 1985).

For the filamentous fungi significant progress has been made with several fine examples of cloning vector development in a number of academically studied species. Unfortunately, such vectors do not have the host range flexibility which makes many of the vehicles available for streptomyces so attractive. Due to this limitation, initially, to clone genes involved in antibiotic biosynthesis from filamentous fungi, alternative strategies to those employed so successfully with Streptomyces have had to be explored. Already the structural gene for isopenicillin N synthetase from Acremonium chrysogenum has been cloned (Sampson et al., 1985). This was achieved in an indirect manner using a synthetic oligonucleotide probe thus circumventing the need for a high frequency transformation vector.

Taken together the achievements witnessed in the Streptomyces and filamentous fungi indicate success can be achieved in the cloning of genes involved in antibiotic biosynthesis. The degree and speed with which such success can be realised is now only limited by the ingenuity and enterprise of the researchers, regardless of the type of producing organism or the antibiotic synthesized.

REFERENCES

Alfoldi, L., 1982, Fusion of microbial protoplasts: problems and perspectives, in: "Genetic engineering of microorganisms for chemicals", A. Hollaender, ed., Plenum Press, New York, pp 59-71.

Altenbuchner, J. and Cullum, J., 1984, DNA amplication and an unstable arginine gene in Streptomyces lividans 66, Mol. Gen. Genet., 195, 1/2:134-138.

Andreoli, P.M., 1985, Versitile E. coli-Bacillus shuttle vectors derived from runaway replication plasmids related to CloDF13, Mol. Gen. Genet., 199:372-380.

Bailey, C.R., Butler, M.J., Normansell, I.D., Rowlands, R.T. and Winstanley, D.J., 1984, Cloning a Streptomyces clavuligerus genetic locus involved in clavulanic acid biosynthesis, Bio/Technology, September, pp 808-811.

Ballance, D.J., Buxton, F.P. and Turner, G., 1983, Transformation of Aspergillus nidulans by orotidine-5'-phosphate decarboxylase gene of Neurospora crassa, Biochem. Biophys. Res. Commun., 112:284-289.

Ballance, D.J. and Turner, G., 1985, Development of a high-frequency transforming vector for Aspergillus nidulans, Gene, 36:321-331.

Banks, G.R., 1983, Transformation of Ustilago maydis by a plasmid containing yeast 2μm DNA, Curr. Genet., 7:73-77.

Berdy, J., 1974, Recent developments of antibiotic research and classification, Adv. Appl. Microbiol., 18:309-406.

Bibb, M.J., Freeman, R.F. and Hopwood, D.A., 1977, Physical and genetical characterization of a second sex factor, SCP2, for Streptomyces coelicolor, MGG, 154:155-166.

Bibb, M.J., Schottel, J.L and Cohen, S.N., 1980, A DNA cloning system for interspecies gene transfer in antibiotic-producing Streptomyces, Nature, 284:526-531.

Bibb, M.J., Ward, J.M., Kieser, T., Cohen, S.N. and Hopwood, D.A., 1981, Excision of chromosomal DNA sequences from Streptomyces coelicolor forms a novel family of plasmids detectable in Streptomyces lividans, Mol. Gen. Genet., 184:230-240.

Bibb, M.J. and Hopwood, D.A., 1981, Genetic studies of the fertility plasmid SCP2 and its SCP2* variants in Streptomyces coelicolor A3(2), J. Gen. Microbiol., 126:427-442.

Bibb, M.J.and Cohen, S.N., 1982, Gene expression in Streptomyces. Construction and application of promoter-probe plasmid vectors in Streptomyces lividans, Mol. Gen. Genet., 187:265-277.

Bibb, M.J., Chater, K.F. and Hopwood, D.A., 1983, Developments in Streptomyces cloning, in: "Experimental manipulation of gene expression", M. Inouye, ed., Academic Press, New York, pp 54-82.

Birch, A.W. and Cullum, J., 1985, Temperature-sensitive mutants of the Streptomyces plasmid pIJ702-characterization; potential application in cloning sust systems and transposon mutagenesis system development, J. Gen. Microbiol., 131(6):1299-303.

Bolotin, A.P., Sorokin, A.V., Aleksandrov, N.N., Danilenko, V.N. and Kozlov, Y.I., 1985, Investigation of Streptomyces plasmid replication: nucleotide sequence of DNA in plasmid pSB24.2- from Streptomyces lividans 66, Antibiot. Med. Biotekhnol, 30(11):804-11.

Bull, A.T., Holt, G. and Lilly, M.D., 1982, "Biotechnology, International Trends and Perspectives", OECD, Paris.

Buxton, F.P., Gwynne, D.I. and Davies, R.W., 1985, Transformation of Aspergillus niger using the argB gene of Aspergillus nidulans, Gene, 37:207-214.

Case, M.E., Schweizer, M., Kushner, S.R. and Giles, N.H., 1979, Efficient transformation of Neurospora crassa by utilising hybrid plasmid DNA, Proc. Natl. Acad. Sci., 76:5259-5263.

Chambers, A.E. and Hunter, I.S., 1984, Construction and use of a bifunctional streptomycete cosmid, Biochem. Soc. Trans., 12:644-64.

Chater, K.F., Hopwood, D.A., Kieser, T. and Thompsn, C.J., 1982, Gene cloning in Streptomyces, Curr. Top. Microbiol. Immunol., 96:69-95.

Chater, K.F., 1984, Morphological and physiological differentiation in Streptomyces, in: "Microbial Development", R. Losick and L. Shapiro, eds., Cold Spring Harbour Laboratory, Cold Spring Harbour, New York, pp 89-115.

Chater, K.F. and Buxton, C.J., 1985, Resistance, regulatory and production genes for the antibiotic methylenomycin are clustered - analysis of antibiotic pathway genes in Streptomyces species, EMBO J., 4(7):1893-97.

Daniel, D. and Tiraby, G., 1982, A Survey of plasmids among natural isolates of Streptomyces, J. Anti., 36:181-183.

Danilenko, V.N., Potekhin, Y.A., Biryukova, I.V., and Navashin, S., 1984, Cloning of DNA in actinomycetes: construction of vector systems, Antibiotiki, 29(8):563-572.

Danilenko, V.N. and Orlova, V.A., 1983, Multiple repetition of the εSA1 element in the genome of Streptomyces antibioticus and its usage for constructing an integrative amplifying vector - improved antibiotic oleandomycin production, Genet. Differentiation Actinomyces, 5.

Demain, A.L., 1983, Metabolic control of secondary biosynthetic pathways, in: "Biochemistry and Genetic Regulation of Commercially Important Antibiotics", L.C. Vining, ed., Biotech. Ser., vol. 2.

Distler, J. and Piepersberg, W., 1985, Cloning and characterization of a gene from Streptomyces griseus coding for a streptomycin-phosphry-lating activity - application of this aph D2 gene in e.g. vector construction, FEMS. Microbiol. Lett., 28, (1):113-17.

Ehrlich, S.D., Jupp, S., Niaudet, B. and Goze, A., 1978, in: "Genetic Engineering", H.W. Boyer and S. Nicosia, eds., Elsevier/North-Holland Publ., Amsterdam, pp 25-32.

Ehrlich, S.D., 1977, Replication and expression of plasmids from Stapphylo-coccus aureus in B. subtilis, Proc. Natl. Acad. Sci. USA, 74:1680-1682.

Ehrlich, S.D., 1978, DNA cloning in B. subtilis, Proc. Natl. Acad. Sci. USA, 75(3):1433-1436.

Elander, R.P., 1982, in: "Trends in Antibiotic Research - Genetics, Biosynthesis, Action and New Substances", H. Umezawa, A.L. Demain, T. Hata and C.R. Hutchinson, eds., Antibiotics Research Assoc., Japan, pp 16-27.

Feitelson, J.S. and Hopwood, D.A., 1983, Cloning of s Streptomyces gene for an O-methyltransferase involved in antibiotic synthesis, Mol. Gen. Genet., 190:394-398.

Fishman, S.E., Rosteck Jr., P.R. and Hershberger, C.L., 1985, A 2.2-kilobase repeated DNA segment is associated with DNA amplification in Streptomyces fradiae - enhanced antibiotic synthesis, J. Bacteriol., 161(1):199-206.

Foor, F., Roberts, G.P., Morin, N., Snyder, L., Hwang, M. and Gibbons, P.H., 1985, Isolation and characterization of the Streptomyces cattleya temperate phage TG1 - for use as cloning vector, Gene, 39(1):11-16.

Friedlin, E. and Nuesch, J., 1984, Isolation of a selective marker from Cephalosporium acremonium by complementation of an auxotrophic mutant of E. coli, Curr. Genet., 8:271-276.

Gil, J.A. and Hopwood, D.A., 1983, Cloning and expression of a p-aminobenzoic acid synthetase gene of the candicidin-producing Streptomyces griseus, Gene, 25:119-132.

Grant, D.M. Lambowitz, A.M. Rambasek, J.A.and Kinsey, J.A., 1984, Transformation of Neurospora crassa with recombinant plasmids containing the cloned glutamate dehydrogenase (am) gene: evidence for autonomous replication of the transforming plasmid, Mol. Cell. Biol., 4(10):2041-2051.

Gray, G., Selzer, G. Buell, G., Shaw, P., Escanez, S. and Thompson C.J., 1984, Synthesis of bovine growth by Streptomyces lividans - new cloning and expression vector, Gene, 32, 1-2:21-30.

Gryczan, T.J.and Dubnau, D., 1978, Construction and properties of chimeric plasmids in B. subtilis, Proc. Natl. Acad. Sci. USA, 75(3):1428-1432.

Herbert, C.J., Sarwar, M., Ner, S.S., Giles, I.G. and Akhtar, M., 1986, Sequence and interspecies transfer of an amino-glycoside-phosphotrans-ferase gene (APH) of B. circulans: self-defence mechanism in antibiotic-producing organisms - butirosin prdoucer, Biochem. J., 233(2):383-393.

Herschberger, C.L., Larson, J.L. and Fishman, S.E., 1983, Uses of recombinant DNA for analyses of Streptomyces species, in: "Biochemical Engineering. III", K. Venkatasubramanian, A. Constantinides and W.R. Vieth, eds., vol. 413, New York Academy of Sciences, pp 31-43.

Hollenberg, C.P., 1982, Cloning with 2µm DNA vectors and the expression of foreign gene in S. cerevisiae, in: "Gene Cloning in Organisms other than E. coli", P.H. Hofschneider and W. Goebel, eds., Springer Verlag, pp 119-144.

Holloway, B.W., Krishnapillai, V. and Morgan, A.F., 1979, Chromosomal genetics of Pseudomonas, Microbiol. Rev., 43:73-102.

Holt, G. and Saunders, G., 1985, Genetic modifications of industrial micro-organisms, in: "Comprehensive Biotechnology", C.L.Cooney and A.E. Humphrey, eds., Pergamon Press.

Hopwood, D.A., Bibb, M.J., Bruton, C.J., Chater, K.F., Feitelson, J.S. and Gil, J.A., 1983, Cloning Streptomyces genes for antibiotic production, Trends in Biotechnology, 1(2):42-48.

Hopwood, D.A., 1985, Genetics and the production of new molecules - antibiotics from Streptomyces, Pestic. Sci., 16(4):425-26.

Hopwood, D.A., Malpartida, F., Keiser, H.M., Ikeda, H., Duncan, J., Kuju, B., Rudd, A., Floss, H.G. and Omura, S., 1985, Production of hybrid antibiotics by genetic engineering, Nature, 314:642-644.

Hopwood, D.A., Keiser, T., Lydiate, D.J. and Bibb, M.J., 1986, Streptomyces plasmids and their biology, in: "Antibiotic Producing Streptomyces", vol. IX, L.E. Day and S.W. Queener, eds., Academic Press, New York, in press.

Horinouchi, S. and Beppu, T., 1984, Production in large quantities in actinorhodin and undecylprodigiosin induced by afsb in Streptomyces lividans - potential use in enhanced antibiotic production, Agric. Biol. Chem., 48(8):2131-2133.

Horinouchi, S. and Beppu, T., 1985, Construction and application of a promoter-probe plasmid that allows chromogenic identification in Streptomyces lividans - application to enchanced expression of antibiotic biosynthetic genes etc., J. Bacteriol., 162(1):406-12.

Horinouchi, S., Nishiyama, M., Suzuki, H., Kumada, Y and Beppu, T., 1985, The cloned Streptomyces bikiniensis A-factor determinant - expression in other Streptomyces spp. and identification of transcription control signal, J. Antibiot., 39(5):636-41.

Hoshino, T., Uozumi, T. and Beppu, T., 1984, Cloning and expression of a gene for resistance to butirosin both in E. coli and B. subtilis - Bacillus vitellinus gene expression in foreign hosts using pTA1302 shuttle vector, Agric. Biol. Chem, 48(2):307-316.

Hosoya, H., Ohtake, Y., Tomizuka, N. and Kurukawa, K., 1984, Construction of a promoter-cloning vector in Pseudomonas aeruginosa, Agric. Biol. Chem., 48(12):3145-3146.

Huetter, R. and Hintermann, G., 1985, Genetic instability in streptomycetes - plasmid and chromosomal instability effect, NATO Adv. Sci. Inst. Ser. Ser. A. (1985), 87, Ind. Asp. Biochem. Genet., pp 27-34.

Izumiya, N., Kato, T., Aoyagi, H., Waki, M. and Kondo, M., 1979, "Synthetic aspects of biologically active cyclic peptides - Gramicidin S and tyrocidines", Kodansha, Tokyo.

Jiménez, A. and Davies, J., 1980, Expression of a transposable antibiotic resistance element in Saccharomyces, Nat., 287:869-871.

Johns, M.A. and Peberdy, J.F., 1984, Transformation of Aspergillus nidulans using the argB gene, Enz. Micro. Technol., 6:386-389.

Jones, M.D. and Fayerman, J.T., 1984, pFJ269: a new plasmid isolated from a beta-lactam antibiotic-producing streptomycete - construction of broad host-range cloning vector for Streptomyces spp., J. Antibiot., 37(12):1727-28.

Jones, G.H. and Hopwood, D.A., 1984a, Activation of phenoxazinone - synthase expression in Streptomyces lividans by cloned DNA sequences from Streptomyces antibioticus - antibiotic actinomycin biosynthetic enzyme, J. Biol. Chem., 259(22):14158-64.

Jones, G.H. and Hopwood, D.A., 1984b, Molecular cloning and expression of the phenoxayinone-synthase gene from Streptomyces antibioticus - cloning and gene expresssion in Streptomyces lividans; antiobiotic actinomycin biosynthesis enzyme, J. Biol. Chem., 259(22):14151-57.

Kato, F., Tanaka, M., Kaito, K., Ishikawa, J. and Koyama, Y., 1985, Studies on the plasmids harbored in Streptomyces roseochromogenus - antibiotic production; cloning vector construction, J. Pharmacobiodyn., 8(2):s-32.

Katz, E. and Demain, A.L., 1977, The peptide antibiotics of Bacillus - chemistry, biogenesis and possible functions, Bact. Rev., 41:449-474.

Katz, E., Thompson, C.J. and Hopwood, D.A., 1983, Cloning and expression of the tyrosinase gene from Streptomyces antibioticus in Streptomyces lividans, J. Gen. Micro., 129:2703-2714.

Kelly, J.M. and Hynes, M.J., 1985, Transformation of Aspergillus niger by the amds gene of Aspergillus nidulans, EMBO J., 4,2:475-479.

Kieselburg, M.K., Weickert, M. and Vary, P.S., 1984, Analysis of resident and transformant plasmids in B. megaterium, Bio/Technology, pp 254-259.

Keiser, T., Hopwood, D.A., Wright, H.M. and Thompson, C.J., 1982, pIJ101, a multicopy broad host-range Streptomyces plasmid: functional analysis and development of DNA cloning vectors, Mol. Gen. Genet., 185:223-238.

Kendall, K. and Cullum, J., 1984, Cloning and expression of an extracellular agarase gene from Streptomyces coelicolor A3(2) in Streptomyces lividans 66 - potential use as protein secretion vector, Gene, 29(3):315-21.

Kintak, K., Ono, H., Tsubotani, S., Harada, S. and Okazaki, H., 1984, Thiotropocin. A new sulphur-containing 7-membered-ring antibiotic produced by a Pseudomonas sp., J. Antibiot., 37(11):1294-1300.

Kirby, R., Wright, L.F. and Hopwood, D.A., 1985, Plasmid-dtermined antibiotic synthesis and resistance in Streptomyces coelicolor, Nature, 254:265-267.

Kirby, R., 1978, An unstable genetic element affecting the production of the antibiotic holomycin by Streptomyces clavuligerus, FEMS Microbiology Lett., 3:283-286.

Kirby, R. and Hopwood, D.A., 1977, Genetic determination of methylenomycin synthesis by the SCP1 plasmid of Streptomyces coelicolor A3(2), J. Gen. Microbiol., 98:239-252.

Kirby, R., Lewis, E. and Botha, C., 1982, A survey of Streptomyces species for covalently closed circular (ccc) DNA using a variety of methods, FEMS Microbiol. Lett., pp 79-82.

Kobayashi, T., Shimotsu, H., Horinouchi, S., Vozumi, T. and Beppu, T., 1984, Isolation and characterisation of a pock forming plasmid pTA40001 from Streptomyces lavendulae, J. Anti., 37(4):368-374.

Kok, M., Eggink, G., Witholt, B., Owen, D.J. and Shapiro, J.A., 1984, Transposable elements as cloning vectors: characterization of the Pseudomonas putida alk-regulon-shuttle vector for alkaline oxidation gene analysis using transposon Tn3-detta-596, Prog. Ind. Microbiol., 20, Innovations in Biotechnol., pp 381-390.

Krause, M., Marahiel, M.A., Dohren, H. and Kleinkauf, H., 1985, Molecular cloning of an ornithine-activating fragment of the Gramicidn S synthetase 2 gene from B. brevis and its expression in E. coli, J. Bact., 162(3):1120-1125.

Kuhstoss, S., Belagaje, R., Hsiung, H. and Rao, R.H., 1985, Characterization of promoters and terminators in Streptomyces ambofaciens - using dual host plasmid pKc356, Abstr. Annu. Meet. Am. Soc. Microbiol., 85 Meet.:139.

Larson, J.L. and Hershberger, C.L., 1984, Shuttle vectors for cloning recombinant DNA in Escherichia coli and Streptomyces griseofulcus C581, J. Bacteriology, 157(1):314-317.

Lilly, E., 1983, Plasmid PEL7 and related cloning vectors for use in Streptomyces, US patent 4416-994.

Majumder, S., Ghosh, S.K., Mukhopadhyay, N.K. and Bose, S.K., 1985, Accumulation of peptides by mycobacillin-negative mutants of B. subtilis B3, J. of Gen. Microbiol., 131:119-127.

Malpartida, F., Zalacain, M., Jiménez, and Davies, J., 1983, Molecular cloning and expressionin Streptomyces lividans of a hygromycin B phosphotransferase gene from Streptomyces hygroscopicus, Biochem. Biophys. Res. Comm., 117(1):6-12.

Manis, J.J. and Clemens, D.L., 1984, Construction and use of a three gene casette for genetic marking of cryptic plasmids in Streptomyces, Abst. Ann. Meet. Am. Soc. Micro., 84 meet.:118.

Martin, J.F. and Gil, J.A., 1984, Cloning and expression of antibiotic production genes, Bio/Technology, January, pp 63-75.

Matsuda, A. and Komatsu, K., 1985, Molecular cloning and structure of the gene for 7β-(4-carboxybutanamido) cephalosporanic acid acylase from a psudomonas strain, J. Bact., 163(3):1222-1228.

Minuth, W., Tudzynski, P. and Esser, K., 1982, Extrachromosomal genetics of Cephalosporium acremonium I. Characterisation and mapping of mitochondrial DNA, Curr. Genet., 5227-231.

Mishra, N.C., 1985, Gene transfer in fungi, Adv. in Genet., 23:74-177.

Morino, T., Isogai, T., Takahasi, H. and Saito, H., 1984, Construction of phage vectors in Streptomyces: introduction of the thiostrepton resistant gene in K4 phage, Agric. Biol. Chem., 48(8):1985-1990.

Murakami, T., Nojiri, C., Toyama, H., Hayashi, E., Katumata, K., Anzai, H., Matsuhasi, Y., Yamada, Y. and Nagaoka, K, 1983, Cloning of antibiotic-resistance genes in Streptomyces, J. Antibiot., 36(10):1305-1311.

McLaughlin, J.R., Murray, C.L. and Rabinowitz, J.C., 1981, J. Biol. Chem., 256:11283-11291.

Nabeshima, S., Hotta, Y. and Okanishi, M., 1984, Construction of plasmid vectors form S. kasugaensis plasmids; pSK1 and pSK2, J. Anti., 37((9):1026-1037.

Nishio, C., Komura, S. and Kurahashi, K., 1983, Peptide antibiotic subtilin is synthesized via precursor proteins - B. subtilus radiolabelling experiments, Biochem. Biophy. Res. Commun., 116(2):751-758.

Normansell, P.J.M., Normansell, I.D. and Holt, G., 1979, Genetic and biochemical studies of mutants of Penicillium chrysogenum impaired in penicillin biosynthesis, J. Gen. Micro., 112:113-126.

Ogata, S., Suenaga, H. and Hayashida, S., 1985, A temperate phage of Streptomyces azureus - isolation and characterization, potential cloning vector for antibiotic(s) producer, Appl. Environ. Microbiol., 49(1):201-04.

Ohnuki, T., Imanaka, T. and Aiba, S., 1985, Self-cloning in Streptomyces griseus of an str gene cluster for streptomycin biosynthesis and streptomycin resistance - strR, strA, strB and strC gene analysis, J. Bacteriol., 164(1):85-94.

Okanishi, M., Kushara, H., Ichihara, M. and Hirasawa, K., 1983, Preparation of a gene bank of Streptomyces kasugaensis by self-cloning, Jap. J. Antibiot., 36(8):2293.

Olsen, R.H., Debusscher, G. and McCombie, W.R., 1982, Development of broad-host-range vectors and gene banks: self cloning of the Pseudomonas aeruginosa PAO chromosome, J. Bact., 150(1):60-69.

Parker, W.L., Rathnum, M.L., Seiner, V., Trejo, W.H., Principe, P.A. and Sykes, R.B., 1984, Cepacin A and Cepacin B. Two new antibiotics produced by Pseudomonas cepacia, J. Antiobiot., 37(5):436-440.

Penalva, M.A., Tourino,A., Sanchez, F., Patino, C., Rubio, U, and Fernandez Sousa, J.M., 1985, Transformation of Acremonium chrysogenum, J. Cell Biochem., Suppl. 9C, p 172.

Perez-Gonzalez, J.A. and Jimenez, A., 1984, Cloning and expression in Streptomyces lividans of a paromomycin-phosphotransferase in Streptomyces simosus forma paramomycinus - paromomycin resistance; dominant selectable marker for cloning vector, Biochem. Biophys. Res. Commun., 125(3):897-901.

Peypoux, F., Marion, D., Maget-Dana, R., Ptak, M., Das, C.B. and Michel, G., 1985, Structure of bacillomycin F, a new peptido-lipid antibiotic of the iturin group - from B. subtilis, Eur. J. Biochem., 153(2):335-340.

Rhodes, P.M., Hunter, I.S., Friend, E.J. and Warren, M., 1984, Recombinant DNA methods for the oxytetracycline producer Streptomyces rimosus, Biochem. Soc. Trans., 12:586-587.

Robbins, W.R., Wirth, D.F. and Hering, C., 1981, Expression of the Streptomyces enzyme endoglycosidase H in E. coli, J.Biol. Chem., 256:10640-10644.

Rodgers, W.H., Springer, W. and Young, F.E., 1982, Cloning and expression of a Streptomyces fradiae neomycin resistance gene in Escherichia coli, Gene, 18:133-141.

Rodicio, M.R., Bruton, C.J. and Chater, K.F., 1985, New derivatives of the Streptomyces temperate phage ØC31 useful for the cloning and functional analysis of Streptomyces DNA - KC500 vector derivatives, Gene, 34(2-3):1137-45.

Roth, M., Guether, R. and Noack, D., 1983, Maintenance of the pIJ2 plasmid in chemostat cultures of Streptomyces lividans 66 (pIJ2) - neomycin phosphotransferase, Genet. Differentiation Actinomycetes, 17.

Richardson, M.A., Mabe, J.A., Beerman, N.E., Nakatsukasa, W.M. and Fayerman, J.T., 1982, Development of cloning vehicles from the Streptomyces plasmid pFJ103, Gene, 20:451-457.

Richardson, M.A., Balagaje, R. and Fayerman, J.T., 1985, Heterologous gene expression in Streptomyces-plasmid vector construction; Bacillus subtilis veg promoter; Streptomyces ambofaciens gene expression, Abstr. Ann. Meet. Am. Soc. Microbiol., 85 Meet.:140.

Rowlands, R.T. and Normansell, I.D., 1983, Current strategies in industrial selection, in: "Bioactive Microbial Products 2", L.J. Nisbet and D.J. Winstanley, eds., Academic Press, pp 1-13.

Samson, S.M., Belaje, R., Blankenship, D.J., Chapman, J.L., Perry, D.A., Skatrud, P.L., Vanfrank, R.M., Abraham, E.P., Baldwin, J.F., Queener, S.W. and Ingolia, T.D., 1985, The isolation, sequence determination and expression in Echerichia coli of the isopenicillin N synthetase gene from Cephalosporium acremonium, Nature, 318:191-192.

Saunders, G., Rogers, M.E., Adlard, M.W. and Holt, G., 1984a, Chromatographic resolution of nucleic acids extracted from Penicillium chrysogenum, Molec. Gen. Gener. 194:343-346.

Saunders, G., Rogers, M.E., Adlard, M.W. and Holt, G., 1984b, Chromatographic resolution of nucleic acids: application to organisms of industrial importance, Trans. Biochem. Soc., 12:694-695.

Saunders, G., Tuite, M.F. and Holt, G., 1986, Gene cloning in Fungi, Trends in Biotech., 4:93-98.

Saunders, G. and Holt, G., 1986, Genetics of the Penicillia, in: "Penicillium and Acremonium", J.F. Peberdy, ed., Biotechnology Handbook Series, Plenum Press, New York, in press.

Seno, E.T.and Chater, K.F., 1983, Glycerol catabolic enzymes and their regulation in wild-type and mutant strains of <u>Streptomyces coelicolor</u> A3(2), J. Gen. Microbiol., 129:1403-1413.

Seno, E.T., Bruton, C.J. and Chater, K.F., 1984, The glycerol utilisation operon of <u>S. coelicolor</u>: genetic mapping of <u>gyl</u> mutations and the analysis of cloned <u>gly</u> DNA, Molec. Gen. Genet., 193(1):119-128.

Sharek, F., Sasarman, A. and Vezina, C., 1984, Construction of a shuttle vector and expression in <u>S. lividans</u> of the sulphonamide resistance gene derived from <u>E. coli</u> plasmid pSAS1206, Can. J. Micro., 30(4):515-518.

Shirai, M., Suzuki, N. and Aida, T., 1984, Effects of colistin on the cells of a colistin-producer - <u>B. colistinus</u> and non-producing and colistin-sensitive strains obtained by mutagenesis, Agric. Biol. Chem., 48(2):521-523.

Shindoh, Y., Nakano, M.M. and Ogawara, H., 1984, Pock forming plasmids isolated from <u>Streptomyces roseochromogenes</u>, J. Anti., 37(5):512-517.

Stahl, U., Tudzynski, P., Kuck, U. and Esser, K., 1982, Replication and expression of a bacterial-mitochondrial hybrid plasmid in the fungus <u>Podospora anserina</u>, Proc. Natl. Acad. Sci., 79:3641-3645.

Stanzak, R., Matsushima, P., Baltz, R.H. and Rao, R.N., 1986, Cloning and expression in <u>S. lividans</u> of clustered erythromycin biosynthesis genes from <u>S. erythreus</u>, Bio/Technology, March, pp 229-232.

Stinchcomb, D.T., Thomas, M., Kelly, J., Selker, E. and Davis, R.W., 1980, Eukaryotic DNA segments capable of autonomous replication in yeast, Proc. Natl. Acad. Sci., 77:4559-4563.

Stohl, L.L. and Lambowitz, A.M., 1983, Construction of a shuttle vector for the filamentous fungus <u>Neurospora crassa</u>, Proc. Natl. Acad. Sci., 80:1058-1062.

Storm, D.R., Rosenthal, K.S. and Swanson, P.E., 1977, Polymyxin and related peptide antibiotics, Ann. Rev. Biochem., 46:723-763.

Sugawara, K., Konishi, M. and Kawaguchi, H., 1984, BMY-28160, a new peptide antibiotic, J. Antibiot., 37(10):1257-1259.

Taylor, D.P., Eckhardt, T. and Fare, L.R., 1983, Gene cloning in the actinomycetes, Ann. N.Y. Acad. Sci., Biochem. Eng. III, 413:47-56.

Thompson, C.J., Ward, J.M. and Hopwood, D.A., 1982, Cloning of antibiotic-resistance and nutritional genes in <u>Streptomyces</u>, J. Bacteriol., 151:668-672.

Thompson, C.J. and Davies, J.E., 1984, Genetic engineering and amino-glycoside antibiotics, Trends in Biotech., 2:43-46.

Thompson, C.J. and Davies, J.E., 1985, Streptomycete plasmid cloning vectors - vector construction and biotechnological application in eg. antibiotics production, NATO Ad. Sci. Inst. Ser. Ser. A. (1985), 87, Ind. Asp. Biochem. Genet., 19-26.

Tichy, P., Smrckova, I., Spizek, J., Rysavy, P. and Kleczek, P., 1984, Cloning of promoter region of aph gene isolated from pIJ2 plasmid DNA from <u>Streptomyces lividans</u> - in <u>Escherichia coli</u>, Folia Microbiol., 29(5):394.

Tilburn, J., Scazzacchio, C., Taylor, G.G., Zabicky-Zissman, J.H., Lockington, R.A. and Davies, R.W., 1983, Transformation by integration in <u>Aspergillus nidulans</u>, Gene, 26:205-221.

Tohyama, H., Shigyo, T. and Okami, Y, 1984, Cloning of steptomycin-gene from a streptomycin-producing streptomycete - cloning of <u>Strepto-</u><u>myces griseus</u> gene in <u>Streptomyces lividans</u>; selectable marker, J. Antibiot., 37(12)1736-37.

Tudzynski, P. and Esser, K., 1982, Extrachromosomal genetics of <u>Cephalo-</u><u>sporium acremonium</u> II. Development of a mitochondrial hybrid vector replicating in <u>S. cerevisiae</u>, Curr. Genet., 6:153-158.

Tudzynski, P. and Esser, K., 1985, Mitochondrial DNA for gene cloning in eukaryotes, <u>in</u>: "Gene Manipulations in Fungi", J.W. Bennett and L.L. Lasure, eds., Academic Press, pp 403-416.

Turgeon, B.G., Garber, R.C. and Yoder, O.C., 1985, Transformation of the fungal maize pathogen Cochliobolus heterostrophus using the Aspergillus nidulans amds gene, Mol. Gen. Genet., 201:450-453.

Turner, G. and Ballance, D.J., 1985, Cloning and transformation in Aspergillus, in: "Gene Manipulations in Fungi", J.W. Bennett and L.L. Lasure, eds., Academic Press, pp 259-278.

Vallins, W.J.S. and Baumberg, S., 1985, Cloning of a DNA fragment from Streptomyces griseus which directs streptomycin-phosphotransferase activity - and is involved in antibiotic resistance and production, J. Gen. Microbiol., 131(7):1657-69.

Van Solingen, P., Muurhing, H.D. and Koekman, B.P., 1985, Transformation of Penicillium chrysogenum using plasmid pGB83, J. Cell Biochem, Suppl. 9c, p 174.

Vapnek, D., Hautala, J.A., Jacobson J.W., Giles, N.H. and Kushner, S.R., 1977, Expression in E. coli of the structural gene for catabolic dehydroquinase of Neurospora crassa, Proc. Natl. Acad. Sci., 74:3508-3512.

Vara, J., Malpartida, F., Hopwood, D.A. and Jimenez, A., 1985, Cloning and expression of a puromycin N-acetyltransferase gene from Streptomyces alboniger in Streptomyces lividans and Escherichia coli - Escherichia coli lac promoter effect; vector selection marker, Gene, 33(2):197-206.

Von Tersch, M.A. and Carlton, B.C., 1983, Megacinogenic plasmids of B. megaterium, J. Bact., 155:872-877.

Ward, M., Wilkinson, B. and Turner, G., 1986, Transformation of Aspergillus nidulans with a cloned, oligomycin-resistant ATP synthase subunit 9 gene, Mol. Gen. Genet., 202:265-270.

Wells, J.S., Trejo, W.H., Principe, P.A. and Sykes, R.B., 1984, Obafluorin: a novel beta-lactone produced by Pseudomonas fluorescens: taxonomy fermentation and biological properties - culture medium, J. Antibiot., 37(7):802-803.

Werneke, J.M., Sligar, S.G. and Schuler, M.A., Development of broad-host-range vectors for expression of cloned genes in Pseudomonas - E. coli shuttle vector, Gene, 38(1-3):73-84.

Westpheling, J., 1983, Host/vector systems for Actinomycetes and applications to strain development, Basic Life Sciences, 25:271-285.

Wood, D.O., Hollinger, M.F. and Tindol, M.B., 1981, Versitile cloning vector for Pseudomonas aeruginosa, J. Bact., 145:1448-1451.

Zimmer, T.L., Frøyshov, Ø and Laland, S.G., 1979, in: "Economic Microbiology Vol. III. Secondary Products of Metabolism", A.H. Rose, ed., Academic Press, New York, pp 123-150.

Zippel, M., Neigenfind, M. and Noack, D., 1983, Possible plasmid involvement in turimycin production in Streptomyces hygroscopicus, Molec. Gen. Genet., 192(3):471-476.

TRANSFER FROM UNIVERSITY TO INDUSTRY

Derek G. Layton

Porton International Limited
29 Chesham Place, London SW1X 8HB, UK

SYNOPSIS

The recent spectacular advances in molecular biology have arisen from basic research, most of which is centrally funded, carried out in university laboratories. Led by the promise of biotechnology's commercial potential and the need for technical expertise industry has been developing closer ties with universities thus intensifying the process of university/industry technology transfer.

Examining university/industry relationships in biotechnology is necessary in order to gain insight into the process of technology transfer and to determine if technology is being transferred in a spirit of cooperation and without compromising the goals of two very different institutions.

*****

Since most of the university/industry relationships in biotechnology are new it is difficult to ascertain how effective they will be in transferring technology between universities and industry. One way of estimating their effectiveness can be made however, by considering the following questions:-

- Why are university/industry relationships in biotechnology being formed?

- Are the relationships working smoothly?

- Has the way research is done in university laboratories, or its quality, been affected by the relationships?

- Has collaboration amongst university researchers been affected?

- Has the quality of education been affected?

- Are there lessons to be learned from previous relationships in such fields as microelectronics?

- What forms are the relationships taking and what are the associated issues?

- Are university policies with respect to the relationships (e.g. patent and royalty agreements, handling of tangible research property, conflicts or interest, etc.) being clearly formulated?

- What is the likely future of and trends in university/industry relationships in biotechnology?

These issues will be examined in this paper with the objective of drawing out guidelines, suggestions for action etc. to improve the efficiency and value of the technology transfer and hopefully ways to reduce the frustrations sometimes felt by both parties. It should be noted however, that the author is subject to bias.

WHY ARE UNIVERSITY/INDUSTRY RELATIONSHIPS IN BIOTECHNOLOGY BEING FORMED?

In general both university and industry representatives agree that universities seeking money from their relationships with industry are motivated in part by a reduction or a fear of reduction in central funding. Industry representatives usually believe, rightly or wrongly, that universities want to gain more "real-world" exposure for faculty and students and are offering them a look at "economic reality". In addition some faculties find industrial funding requires 'less administrative work and is longer term than Government funded renewable grants'!

ARE THE RELATIONSHIPS WORKING SMOOTHLY?

The perception of many is that these relationships are working well. The initial administration of agreements in biotechnology was often inefficient because new policies were being formulated and new players (biologists, instead of engineers), were involved in the interactions. In addition research administrators have had to learn how to administer technology transfer agreements. It may be that the agreements are currently working well because there are, as yet, almost no biotechnology products. Disagreements may arise, especially in limited partnerships, when product sales revenues are generated.

HAS THE WAY RESEARCH IS DONE IN UNIVERSITY LABORATORIES OR ITS QUALITY BEEN AFFECTED BY THE RELATIONSHIPS?

University staff in the main believe that university/industry relationships in biotechnology have had no effect on the way research is done or on its quality. Industrialists are similarly of the belief or illusion that certainly the way in which research is done has been altered, for the better, by those relationships.

HAS COLLABORATION AMONGST UNIVERSITY RESEARCHERS BEEN AFFECTED?

Most university staff believe that such relationships have had no substantial effect on the exchange of information or the collaboration that existed. In fact there is probably only limited collaboration anyway in rapidly evolving areas of science. Most industry representatives are concerned that faculty members having consulting or research agreements keep proprietary information confidential. This concern may reflect a previous lack of clear communication as to the parties' particular responsibilities and of having properly explored current obligations, intentions, etc.

HAS THE QUALITY OF EDUCATION BEEN UNAFFECTED?

It is unlikely that any university would believe that there has been a reduction in the quality of education the students receive since it is the goal of faculty and university administrators to protect and maintain standards of academic excellence. The student body may feel a need to be involved in monitoring standards since funding reductions may provoke disproportionate efforts to secure industrial funds.

ARE THERE LESSONS TO BE LEARNED FROM PREVIOUS RELATIONSHIPS IN SUCH FIELDS AS MICROELECTRONICS?

The development of the semiconductor industry is suggested as a comparison for the development of biotechnology. Virtually all of the basic research in electronic engineering carried out during the 1950's and 1960's was supported by central funding plus some additional specialist defence funding. This facilitated research, training infrastructure and generated increasingly cooperative ties between universities and private companies. This was especially so in the USA. However, whilst this created a dynamic industrial situation few of the actual semi-conductor innovations emerged directly from centrally funded university research.

The potential industrial applications of biotechnology by contrast have emerged directly from publicly funded academic biomedical research. As biotechnology has been moving to the market, universities have been the buffers in commercialising the fruits of public funding because often they are the sole source of the know-how. Many of the new firms in biotechnology have sprung out of academia; whereas in the semi-conductor field government procurement helped to create industrial know-how and encouraged industrial spin-off. The traditional roles of university and industry have become blurred in biotechnology.

WHAT FORMS ARE THE RELATIONSHIPS TAKING AND WHAT ARE THE ASSOCIATED ISSUES?

Five broad types of university/industry arrangements in biotechnology are considered:

- consulting arrangements
- industrial associates programmes
- research contracts
- research partnerships
- private corporations

Consulting Arrangements

Consulting is important for several reasons. It allows direct technology transfer between universities and industry that goes both ways. Academics agree that consulting keeps them appraised of new innovations in industrial R & D and that their knowledge can be applied to new kinds of problems related to, but outside of, their on-campus research. Industry views consulting arrangements with university faculty essentially as having an expert on retainer.

Many universities have policies on consulting but the policies vary. Typically they have, or <u>should have</u>, provisions regarding conflict of interest, time regulation, disclosure and policy enforcement. In most cases, policy enforcement is based on an honour system. Disclosure policies are of interest for public access to objective scientific information; they often cause the greatest concern to both sides.

Industrial Associates Programmes

These usually involve entire university departments or a group of specialists within a department. Companies pay a set annual fee that allows them to participate in seminars, interact with research students and staff and previous publications. Industrial associate programmes are more common in USA. They are seen to facilitate technology transfer between the parties, open up opportunities for further consulting and contract arrangements, provide funding for research and give industry access to graduate students for future employment. Industrialists generally view them as useful, however outsiders of the 'club' often perceive them as selling research on an exclusive basis to 'members'. In fact 'exclusivity' is rarely the basis or purpose of these programmes.

Research Contracts

University research contracts with industrial sponsors have been and continue to be an important type of university/industry relationship in biotechnology. They differ from consulting arrangements in that the industrial sponsor is usually paying for a specific piece of research or supporting general research activities. Contractual arrangements often grow out of consulting and are usually motivated by industry's needs for research that complements research being done in-house or for some expertise in a new area.

Several of the university research contracts with industrial sponsors in biotechnology have been large and have elicited questions regarding issues such as mixing of funds, patent rights, and disclosure of equity or of other financial arrangements between the sponsor and the principal investigator.

Issues of conflict of interest, invention rights, mixing of funds and university policies regarding the processing of contractual arrangements are all important. It is interesting to note the MIT, which traditionally has had a close relationship with industry, has the most explicit guidelines for consulting, disclosure and processing of industry sponsored contracts.

Even with explicit guidelines research contracts may, and often do, run into serious operational problems. These usually arise because university researchers traditionally do not set time bench-marks in research projects and may not understand the basis for commercial evaluation of the results and hence the directional decisions that have to be made.

Research Partnerships

Another type of university/industry arrangement taking place in biotechnology is the joint establishment of a research foundation, institute or long term collaborative arrangement by an industrial sponsor and a university. Examples are Hoechst/Massachusetts General; Monsanto/ Washington and to some extent Leicester/ICI. To date both parties in such relationships appear to be satisfied with them, though the problems which may/will arise will mainly have to be dealt with by the universities; they include mixing of funds where faculty cooperate with others who have central

funding, terms of termination, rights to publish and the external peer
review of projects.

Private Corporations

Innovative approaches to connecting university research to commercial
development in biotechnology are being initiated. The establishment of
Engenics (Stanford and UC Berkeley), Neogen (Michigan State) and Imperial
Biotechnology (I.C. London) are examples of very different approaches.

Engenics was funded by six corporations via a non-profit Centre for
Biotechnology at Stanford which in turn funds campus based research,
royalties from subsequent licences returning to the Centre. Neogen was
established to utilize limited partnerships and tax benefits to stimulate
the progress of entrepreneurial ideas from the faculty, royalties again
funnelled back into a non-profit foundation.

Imperial Biotechnology was launched in order to exploit the College's
pilot plant which though built in the 1960's was in good condition and had
been underused. A first major contract was for the scale-up of Biogen's
interferon process.

ARE UNIVERSITY POLICIES WITH RESPECT TO THE RELATIONSHIPS BEING FORMULATED?

In general at the current time insufficient clear guidelines and
policies have been drawn up by the universities. The filing of patents,
their assignment, the granting of exclusive or non-exclusive licences all
need to be subjects of clear policy statements of the university, even if,
as in the case of licences, the operating is to be decided on a case-by-case
basis. Industry requires for its own planning and risk assessment
guidelines to work with. Similarly policies must be drawn up with regard to
tangible research property and its transfer to the sponsor, and to conflict
of interest, disclosure and resolution of disputes. The mixing of central
and industrial funds is to an extent inevitable no matter how strict the
guidelines are, nevertheless provisions must be made to minimize potential
disputes.

WHAT IS THE LIKELY FUTURE OF AND TRENDS IN UNIVERSITY/INDUSTRY RELATIONSHIPS IN BIOTECHNOLOGY?

University/industry relationships in biotechnology will most likely
follow the same pattern that they have in other high technology areas.
First, scientific break-throughs generate a period of hyperactivity in
university/industry relationships. This hyperactivity phase is
characterised by the promise of "big money" which leads to a short-term
staff and post-graduate drain. After the industry goes through its initial
phases, an equilibrium state is reached and a fairly healthy symbiotic
relationship emerges.

The future of the relationship may well depend largely on the success
of biotechnology companies in getting products into the market in acceptable
timescales. Failures may create disillusionment for both parties especially
for some corporations who may feel that they were "led on" by academics'
promises or that they were sucked into the hyperactivity. Biotechnology
may, therefore, take somewhat longer to become a mature industry than others
in the past. There is also a danger that the anxiety of industry to enter a
particular sector of technology and of university departments to secure
research funds during the current period or erosion of central funds may

compound together and produce a situation in the next decade where
university departments are perceived to have "nothing to offer". Industry
currently expresses such concern over chemical engineering research.

RECOMMENDATIONS

Since university/industry relationships are currently considered to be
desirable by both parties and are not thought to affect either the
educational or fundamental research roles of the university, only
suggestions to ensure the smooth operation of these relations will be made.

The following are some of the 'essential' requirements:

1.   That both parties clearly understand each other's objectives and
      methods of working and accept them prior to entering into agreements.

2.   That the universities set out clear policies and guidelines on
      consulting, conflicts of interest, patents, tangible research property,
      etc. and that they adhere to them.

3.   Industry should be open and straightforward with regard to likely
      royalty incomes thereby trying to create a spirit of commercial
      partnership.

4.   University researchers must understand that industry needs to control
      and plan to benchmarks their research effort. They should seek
      industry's assistance in such planning which in turn should be
      provided.

5.   Changes in research emerging from commercial inputs/decisions should be
      clearly explained to the academic staff.

6.   Any proposals submitted 'cold' to industry by university must be
      professionally presented.

7.   University research staff must accept, if they want the funds, that
      profit is the motivating force for industry and that in the commercial
      environment everything is accountable for.

CONTRIBUTORS

T. BEPPU
Department of Agricultural
  Chemistry
The University of Tokyo
Bunkyo-ku
Tokyo 113
Japan

J.R. BIRCH
Celltech Limited
244-250 Bath Road
Slough SL1 4DY
Berks
England

D.R. BOONE
Division of Environmental and
  Occupational Health Sciences
School of Public Health
University of California
Los Angeles, CA 90024
United States

R. BORASTON
Celltech Limited
244-250 Bath Road
Slough SL1 4DY
Berks
United Kingdom

A.T. BULL
The Institute for Biotechnological
  Studies
Biological Laboratory
The University of Kent
Canterbury
Kent CT2 7NJ
United Kingdom

M.F. CARDOSI
Biotechnology Centre
Cranfield Institute of Technology
Cranfield
Bedford MK43 OAL
England

K.M. DOMBEK
Department of Microbiology and
  Cell Science and Department
  of Immunology and Medical
  Microbiology
1052 McCarty Hall
University of Florida
Gainesville, Florida 32611
United States

J.C. DUARTE
LNETI
2745 Queluz-de-Baixo
Portugal

P. FORD
The School of Biotechnology
The Polytechnic of Central London
115 New Cavendish Street
London W1M 8JS
United Kingdom

Y. FUJIMOTO
School of Pharmacy
University of Wisconsin
Madison
Wisconsin 53706
United States

S. GARLAND
Celltech Limited
244-250 Bath Road
Slough SL1 4DY
Berks
United Kingdom

J.M. GHUYSEN
Service de Microbiologie
Faculté de Médecine
Université de Liège
Institut de Chimie, B6
B-4000 Sart Tilman (Liège)
Belgium

I.J. HIGGINS
Biotechnology Centre
Cranfield Institute of Technology
Cranfield
Bedford MK43 OAL
United Kingdom

G. HOLT
The School of Biotechnology
The Polytechnic of Central London
115 New Cavendish Street
London W1M 8JS
United Kingdom

S. HORINOUCHI
Department of Agricultural
  Chemistry
The University of Tokyo
Bunkyo-ku
Tokyo 113
Japan

Z. IKRAM
The School of Biotechnology
The Polytechnic of Central London
115 New Cavendish Street
London W1M 8JS
United Kingdom

L.O. INGRAM
Department of Microbiology and Cell
  Science and Department of Immuno-
  logy and Medical Microbiology
1052 McCarty Hall
University of Florida
Gainesville, Florida 32611
United States

A.C. KENNEY
Celltech Limited
244-250 Bath Road
Slough SL1 4DY
Berks
United Kingdom

K. LAMBERT
Celltech Limited
244-250 Bath Road
Slough Sl1 4DY
Berks
United Kingdom

D.G. LAYTON
Porton International plc
29 Chesham Place
London SW1M 8HB
United Kingdom

R.A. MAH
Division of Environmental and
  Occupational Health Sciences
School of Public Health
University of California
Los Angeles, CA 90024
United States

M.A. MARTINS-LOUCAO
Department of Plant Biology
Faculty of Sciences
1294 Lisboa Codex
Portugal

J.M. NOVAIS
Laboratório de Engenharia
  Bioquímica
Instituto Superior Técnico
1000 Lisboa
Portugal

V. PETIARD
L.E.R. Synthelabo
Tours
France

T.N. PICKNETT
The School of Biotechnology
The Polytechnic of Central London
115 New Cavendish Street
London W1M 8JS
United Kingdom

C. RODRIGUEZ-BARRUECO
Unit of Nitrogen Fixation
Centro de Edafologia y Biologia
  Aplicada
CSIC
Salamanca
Spain

G. SAUNDERS
The School of Biotechnology
The Polytechnic of Central London
115 New Cavendish Street
London W1M 8JS
United Kingdom

C.J. SIH
School of Pharmacy
University of Wisconsin
Madison
Wisconsin 53706
United States

P. STECK
Sanolfi/Elf Recherche
Toulouse
France

P.W. THOMPSON
Celltech Limited
244-250 Bath Road
Slough SL1 4DY
Berks
United Kingdom

A.P.F. TURNER
Biotechnology Centre
Cranfield Institute of Technology
Cranfield
Bedford MK43 OAL
United Kingdom

S.-H. WU
School of Pharmacy
University of Wisconsin
Madison
Wisconsin 53706
United States

202